VITAMINS AND HORMONES
VOLUME 52

Editorial Board

FRANK CHYTIL

MARY F. DALLMAN

JENNY P. GLUSKER

ANTHONY R. MEANS

BERT W. O'MALLEY

VERN L. SCHRAMM

MICHAEL SPORN

ARMEN H. TASHJIAN, JR.

VITAMINS AND HORMONES
ADVANCES IN RESEARCH AND APPLICATIONS

Editor-in-Chief

GERALD LITWACK

Department of Pharmacology
Jefferson Cancer Institute
Thomas Jefferson University Medical College
Philadelphia, Pennsylvania

Volume 52

ACADEMIC PRESS

San Diego   London   Boston
New York   Sydney   Tokyo   Toronto

This book is printed on acid-free paper. ∞

Copyright © 1996 by ACADEMIC PRESS

All Rights Reserved.
No part of this publication may be reproduced or transmitted in any form or by any means, electronic or mechanical, including photocopy, recording, or any information storage and retrieval system, without permission in writing from the publisher.

Academic Press, Inc.
525 B Street, Suite 1900, San Diego, California 92101-4495, USA
http://www.apnet.com

Academic Press Limited
24-28 Oval Road, London NW1 7DX, UK
http://www.hbuk.co.uk/ap/

International Standard Serial Number: 0083-6729

International Standard Book Number: 0-12-709852-6

PRINTED IN THE UNITED STATES OF AMERICA
96 97 98 99 00 01 BC 9 8 7 6 5 4 3 2 1

Former Editors

ROBERT S. HARRIS
Newton, Massachusetts

JOHN A. LORRAINE
University of Edinburgh
Edinburgh, Scotland

PAUL L. MUNSON
University of North Carolina
Chapel Hill, North Carolina

JOHN GLOVER
University of Liverpool
Liverpool, England

GERALD D. AURBACH
Metabolic Diseases Branch
National Institute of Diabetes
and Digestive and Kidney Diseases
National Institutes of Health
Bethesda, Maryland

KENNETH V. THIMANN
University of California
Santa Cruz, California

IRA G. WOOL
University of Chicago
Chicago, Illinois

EGON DICZFALUSY
Karolinska Sjukhuset
Stockholm, Sweden

ROBERT OLSON
School of Medicine
State University of New York
at Stony Brook
Stony Brook, New York

DONALD B. MCCORMICK
Department of Biochemistry
Emory University School of Medicine
Atlanta, Georgia

Contents

PREFACE xi

Vitamins C and E and LDL Oxidation
Balz Frei, John F. Keaney, Jr., Karen L. Retsky, and Kent Chen

I.	Introduction	1
II.	Mechanisms of LDL Oxidation	5
III.	Antioxidant Protection of LDL by Vitamin C	9
IV.	LDL Oxidation and Vitamin E	14
V.	Trials of *in Vivo* Antioxidant Supplementation	21
VI.	Conclusions	26
	References	27

Antioxidant Vitamins and Human Immune Responses
Adrianne Bendich

I.	Introduction	35
II.	Free Radicals	36
III.	Essential Micronutrients with Antioxidant Activities	36
IV.	Immune Functions	37
V.	Free Radicals and Immune Cell Function	39
VI.	Clinical Examples of the Interactions of Free Radicals, Antioxidants, and Immune Function	42
VII.	Conclusions	55
	References	56

Cytokine Regulation of Bone Cell Differentiation
Melissa Alsina, Theresa A. Guise, and G. David Roodman

I.	Introduction	63
II.	Osteoblasts	63
III.	Osteoclasts	76
IV.	Summary	86
	References	86

The Molecular Pharmacology of Ovarian Steroid Receptors

ELISABETTA VEGETO, BRANDEE L. WAGNER, MARKUS O. IMHOF, AND DONALD P. MCDONNELL

I.	Introduction	99
II.	The Mechanism of Action of Estrogen and Progesterone	101
III.	Steroid Hormone Receptor Antagonists	109
IV.	Final Comments	120
	References	121

Signal Transduction Pathways Combining Peptide Hormones and Steroidogenesis

MICHAEL R. WATERMAN AND DIANE S. KEENEY

I.	Introduction	129
II.	Acute Action of Peptide Hormones on Steroidogenesis	130
III.	Chronic Action of Peptide Hormones on Steroidogenesis	134
IV.	Conclusions and Future Directions	143
	References	145

The Roles of 14-3-3 Proteins in Signal Transduction

GARY W. REUTHER AND ANN MARIE PENDERGAST

I.	Introduction	149
II.	14-3-3 Sequence and Structure Analysis	150
III.	Regulation of Cellular Processes by 14-3-3 Proteins	153
IV.	14-3-3 Proteins and Protein Kinases	158
V.	Roles of 14-3-3 Proteins in Signal Transduction Pathways	166
VI.	Conclusions	169
	References	170

Physiological Roles for Parathyroid Hormone-Related Protein: Lessons from Gene Knockout Mice

ANDREW C. KARAPLIS AND HENRY M. KRONENBERG

I.	Introduction	177
II.	Strategy for Generating PTHrP-Negative Mice	179

III.	The PTHrP-Negative Phenotype	181
IV.	Summary	190
	References	191

INDEX 195

Preface

This volume is an eclectic coverage of topics in the areas of vitamins, cytokines, steroid hormones, specific proteins, and peptide hormones.

In the first category, there are two contributions on vitamins, one from the Balz Frei laboratory on vitamins C and E and LDL oxidation and a second from Adrianne Bendich on antioxidant vitamins and human immune responses. These are followed by one paper on cytokines from the G. David Roodman Laboratory on cytokine regulation of bone cell differentiation. Two chapters in the steroid hormone area follow. The first of these is from the Donald McDonnell laboratory on the molecular pharmacology of ovarian steroid receptors and the second is from Michael R. Waterman's laboratory on signal transduction pathways combining peptide hormones and steroidogenesis. Two papers on specific proteins follow. One of these is from Ann Marie Pendergast's laboratory on the roles of 14-3-3 proteins in signal transduction, and the other is from Andrew C. Karaplis and Henry M. Kronenberg on the physiological roles for parathyroid hormone-related protein: lessons from gene knockout mice.

Certain members of the Editorial Board have suggested some of the topics and authors, for which I am grateful. Academic Press continues as a facilitating partner in the production of this serial.

GERALD LITWACK

Vitamins C and E and LDL Oxidation

BALZ FREI, JOHN F. KEANEY, JR., KAREN L. RETSKY, AND KENT CHEN

Whitaker Cardiovascular Institute, Boston University School of Medicine, Boston, Massachusetts 02118

I. Introduction
  A. The Pathogenesis of Atherosclerosis
  B. The Role of Oxidized LDL in Atherosclerosis
II. Mechanisms of LDL Oxidation
III. Antioxidant Protection of LDL by Vitamin C
  A. Molecular Mechanisms
IV. LDL Oxidation and Vitamin E
  A. Antioxidant Protection of LDL
  B. *In Vitro* Stimulation of LDL Oxidation
V. Trials of *in Vivo* Antioxidant Supplementation
  A. Vitamin C
  B. Vitamin E
VI. Conclusions
  References

I. Introduction

A. The Pathogenesis of Atherosclerosis

Atherosclerosis is a systemic disease of large and medium-sized arteries characterized by hardening and loss of elasticity of the arterial wall and narrowing of the lumen of the artery. The narrowing is mainly due to local thickening of the intima, which is the innermost layer of the arterial wall (i.e., the layer that faces the bloodstream). The principal clinical manifestations of atherosclerosis are myocardial infarction and ischemic stroke, which together cause more than 40% of all deaths in Western societies (Gotto and Farmer, 1988). In the initial stages of atherosclerosis, low-density lipoprotein (LDL) and other lipoproteins, such as lipoprotein (a), accumulate in the subendothelial space of lesion-prone arterial sites (Table I). The endothelium over these sites is activated or dysfunctional, expressing monocyte adhesion molecules and monocyte chemoattractants. As a consequence, monocytes adhere to the endothelium and migrate across the endothelial monolayer to the subendothelial space of the intima. There, the monocytes become activated and differentiate into resident macrophages.

TABLE I
KEY SEQUENTIAL DETERMINANTS OF EARLY ATHEROSCLEROTIC LESION DEVELOPMENT[a]

1. A favorable local hemodynamic environment (lesion-prone sites): domains of low shear stress (entrance regions of arteries and lateral leading edges of flow dividers and orifices).
2. Enhanced focal intimal influx and accumulation of plasma lipoproteins: LDL, small dense LDL (LDL-B), and lipoprotein (a).
3. Augmented net intimal oxidative stress status.
4. Minimal oxidative modification of intimal lipoproteins.
5. Focal blood monocyte recruitment to the arterial intima.
6. Intimal monocyte–macrophage differentiation and activation.
7. Oxidative modification of intimal lipoproteins, partly stimulated by macrophages.
8. Intimal foam cell formation: uptake of oxidized and aggregated LDL by macrophages.

[a]Adapted from Schwartz and Valente (1994).

The early atherosclerotic lesion, called a fatty streak, is characterized by the presence of lipid-laden foam cells, most of which arise from macrophages through massive intracellular lipid accumulation. Fatty streaks may progress via the intermediate lesion to the fibrous plaque (Schwartz and Valente, 1994). Two cardinal events in the transition from the potentially reversible fatty streak to the less reversible intermediate or transitional lesion are (i) foam cell necrosis, resulting in deposition of extracellular cell debris, lipid, and cholesterol; and (ii) smooth muscle cell migration from the media (the middle layer of the arterial wall) to the intima, where the smooth muscle cells proliferate and change from a contractile to a secretory phenotype. These myointimal cells secrete collagen and other extracellular matrix proteins, which results in formation of a dense connective tissue cap covering the atheromatous cellular components (foam cells, macrophages, smooth muscle cells, etc.) and the necrotic lipid core (cell debris, cholesterol crystals, etc.). Finally, mature fibrous plaques are formed by continued foam cell formation and necrosis, continued proliferation of myointimal cells, fibrosis, mural thrombosis (blot clot formation in the arterial wall), and calcification (Schwartz and Valente, 1994). Plaque rupture and occlusive thrombosis may then result in a myocardial infarction or ischemic stroke.

B. THE ROLE OF OXIDIZED LDL IN ATHEROSCLEROSIS

In the current understanding of atherogenesis, LDL itself is not considered atherogenic. Only after LDL has been modified by oxidation does the lipoprotein contribute to lesion formation (Table I). This "oxidative modification hypothesis of atherosclerosis" originated with the seminal observation by Brown and Goldstein more than 15 years ago that macrophages in culture cannot be converted to lipid-laden cells in the presence of native LDL; in order to stimulate macrophage uptake, chemical modification of LDL is required (Goldstein et al., 1979). Henriksen et al. (1981) subsequently recognized that oxidation was one chemical modification that converted LDL into a form rapidly internalized by macrophages. The uptake of oxidatively modified LDL (oxLDL) by macrophages is mediated by scavenger receptors that recognize the modified apolipoprotein B-100 (apo B) moiety of oxLDL. Originally, two types of scavenger receptors (types I and II) were identified, both of which have been cloned and characterized (Krieger and Herz, 1994). Several other receptors that recognize oxLDL have been described (Stanton et al., 1992; Endemann et al., 1993; Ottnad et al., 1995). Unlike uptake of native LDL by the normal apo B/E receptor (present, e.g., on hepatocytes), uptake of oxLDL by macrophages via the scavenger receptor pathway is not subject to downregulation by intracellular cholesterol content. Consequently, macrophages in the intimal subendothelium become loaded with oxLDL-derived lipids and thus are converted to foam cells, which are characteristic of the fatty streak (Schwartz and Valente, 1994) (see earlier).

In addition to increased uptake by macrophages causing foam cell formation, oxLDL may be atherogenic by numerous other mechanisms (Berliner and Heinecke, 1996):

1. *Monocyte–macrophage recruitment and retention.* Mildly oxidized LDL, often referred to as minimally modified LDL, causes increased endothelial adhesiveness toward monocytes (Berliner et al., 1990; Kim et al., 1994). OxLDL also stimulates monocyte–endothelium interactions (Frostegard (et al., 1991), possibly due to upregulation of intercellular adhesion molecule-1 on endothelial cells (Sugiyama (et al., 1994). Furthermore, minimally modified LDL activates endothelial cells to synthesize monocyte chemotactic peptide-1 (MCP-1) and monocyte colony-stimulating factor (M-CSF) via activation of endothelial adenylate cyclase and the transcription factor nuclear factor-κB (Parhami et al., 1993). MCP-1, together with monocyte adhesion molecules, is thought to play an important role in the attraction and guided migration of monocytes into the arterial intima, while M-CSF stimulates the differentiation of monocytes into macrophages. OxLDL itself, probably due to its increased content of lysophosphatidylcholine, is chemotactic for monocytes (Quinn et al., 1987, 1988), which may further facilitate recruitment of these cells into the atherosclerotic lesion.

Finally, oxLDL inhibits migration of resident tissue macrophages and may trap monocyte-derived macrophages in the intima (Quinn et al., 1987).

2. *Foam cell formation by scavenger receptor-independent mechanisms.* OxLDL is immunogenic, stimulating formation of autoantibodies (Palinski et al., 1989). Immune complexes of oxLDL aggregates have been shown to be efficiently internalized by macrophages via F_c receptors (Khoo et al., 1992). In addition, oxLDL can form aggregates that are internalized by macrophages via apoB/E receptor–dependent phagocytosis, leading to massive intracellular lipid deposition (Suits et al., 1989).

3. *Regulation of vascular tone.* OxLDL inhibits the ability of endothelial cells to release endothelium-derived relaxing factor (EDRF) in response to acetylcholine and other receptor-mediated stimuli. EDRF has been identified as nitric oxide (NO). The inhibition of endothelial NO production by oxLDL may be explained by oxLDL's increased content of lysophosphatidylcholine and interruption of G-protein-dependent signal transduction (Flavahan, 1992). OxLDL may also inactivate EDRF after its release from endothelium. EDRF not only acts as a vasodilator but inhibits platelet aggregation, smooth muscle cell proliferation, and endothelium–leukocyte interactions (Welch and Loscalzo, 1994). Therefore, inhibition of EDRF production and action by oxLDL could have widespread consequences for atherosclerotic lesion formation and development of myocardial infarction and ischemic stroke. Finally, oxLDL stimulates production of the vasoconstrictor molecule endothelin by endothelial cells (Boulanger et al., 1992), and contains increased amounts of free radical-derived prostaglandin F_2-like compounds (F_2-isoprostanes), which can act as vasoconstrictors (Lynch et al., 1994).

4. *Cytotoxicity.* OxLDL is cytotoxic by virtue of lipid hydroperoxide breakdown products (Benedetti et al., 1980) and specific cholesterol oxidation products that can initiate lipid peroxidation in target cells (Coffey et al., 1995). Cytotoxicity to foam cells may be particularly relevant to lesion progression (see earlier), while physical damage and denudation of vascular endothelium may contribute to the acute thrombotic complications of atherosclerosis.

5. *Thrombosis and fibrinolysis.* Minimally modified LDL and oxLDL can stimulate synthesis of procoagulant tissue factor by endothelial cells (Drake et al., 1991; Weis et al., 1991). In addition, oxLDL can abolish the ability of endothelial cells to produce the potent inhibitor of platelet thrombus formation, prostacyclin (Thorin et al., 1994), and can cause both platelet activation and aggregation (Aviram, 1989;

Naseem *et al.*, 1993). OxLDL also may impede fibrinolysis by stimulating expression of plasminogen activator inhibitor-1 by endothelial cells (Latron *et al.*, 1991).

By the above mechanisms, oxLDL may play a pivotal role in atherosclerosis, from its initial stages, characterized by monocyte recruitment and foam cell formation, to its clinical manifestations many decades later, such as angina pectoris, myocardial infarction, and ischemic stroke due to impairment of vascular reactivity, vasoconstriction, plaque rupture, and occlusive thrombosis. The prevention of LDL oxidation, therefore, may have significant consequences for the prevention or inhibition of atherosclerotic vascular disease and its clinical sequelae. It is the objective of this article to review the evidence for inhibition of LDL oxidation by the antioxidant vitamins C and E.

II. Mechanisms of LDL Oxidation

In order to understand the mechanisms by which vitamins C and E may affect LDL oxidation, it is necessary to consider the molecular mechanisms of LDL oxidation. While it remains unclear how LDL is modified *in vivo*, the mechanism of oxidative modification of LDL *in vitro* has been studied extensively. Human LDL consists of a lipophilic core containing cholesterol esters and some triglycerides, an outer monolayer of free cholesterol and phospholipids, and one molecule of apo B, which embraces the entire LDL particle (Esterbauer *et al.*, 1990). An average human LDL particle contains about 2700 esterified fatty acids, about half of which are polyunsaturated (Esterbauer *et al.*, 1987). In addition, LDL contains a number of lipid-soluble antioxidants (Table II). The most abundant LDL-associated antioxidant is α-tocopherol, which is the most active form of vitamin E. Each LDL particle contains on average five to nine molecules of α-tocopherol (Esterbauer *et al.*, 1992). Other antioxidants associated with LDL are γ-tocopherol, ubiquinol-10, and several carotenoids and oxycarotenoids (Table II) (Esterbauer *et al.*, 1992; Keaney and Frei, 1994). The content of both polyunsaturated fatty acids (PUFAs) and antioxidants in LDL can vary considerably among individuals as a result of several factors, such as dietary intake, intake of fat, and rate of absorption and utilization.

In vitro studies have determined that the oxidation of human LDL proceeds through a series of events involving lipid peroxidation and covalent modification of apo B. Several stages of LDL oxidation can be

TABLE II
Antioxidant Content of Human LDL

Antioxidant	nmol/mg LDL protein (mean ± SD)	mol/mol LDL (mean)
Vitamin E (α- and γ-tocopherol)	15.5 ± 2.9[a]	7.95
Ubiquinol-10	0.65 ± 0.28[a]	0.33
β-Carotene	0.53 ± 0.47[b]	0.27
Lycopene	0.41 ± 0.20[a,b]	0.21
Cryptoxanthine	0.25 ± 0.23[b]	0.13
α-Carotene	0.22 ± 0.25[b]	0.11

[a]Data from Frei and Gaziano (1993).
[b]Data from Esterbauer et al. (1992).

distinguished: (i) initiation of lipid peroxidation and consumption of LDL-associated antioxidants; (ii) propagation of lipid peroxidation; (iii) decomposition of lipid hydroperoxides to reactive aldehydes; and (iv) covalent modification of apo B by these aldehydes, leading to recognition of LDL by the macrophage scavenger receptors. These stages of LDL modification should not be viewed as separate events, but rather as a continuum of complex and changing interactions between LDL-associated antioxidants, lipids, and apo B.

The earliest event in the modification of LDL is initiation of lipid peroxidation. While abstraction of a doubly allylic hydrogen (a hydrogen on a carbon located between two double bonds) from a PUFA as explained in Fig. 1 represents true *(de novo)* initiation of lipid peroxidation, this is not the prevailing mechanism of initiation of lipid peroxidation in LDL *in vitro*. Because LDL is highly prone to autooxidation, LDL isolation from blood is often associated with *ex vivo* formation of small amounts of lipid hydroperoxides. Upon incubation of isolated LDL, these preformed lipid hydroperoxides break down to alkoxyl radicals, which can initiate lipid peroxidation (see later, Reactions 1–5). The latter process, therefore, should be termed more correctly "reinitiation"; *de novo* initiation of lipid peroxidation by hydrogen abstraction from a PUFA can only be studied in LDL free of detectable amounts of preformed lipid hydroperoxides (Lynch and Frei, 1993, 1995; Ingold et al., 1993).

While the initiating species of LDL lipid peroxidation *in vivo* is unknown, many oxidizing conditions have been utilized *in vitro*. Cultured vascular cells, i.e., endothelial cells (Henriksen et al., 1981; Steinbrecher et al., 1984; van Hinsbergh et al., 1986; Steinbrecher, 1988), smooth muscle cells (Morel et al., 1984; Heinecke et al., 1986),

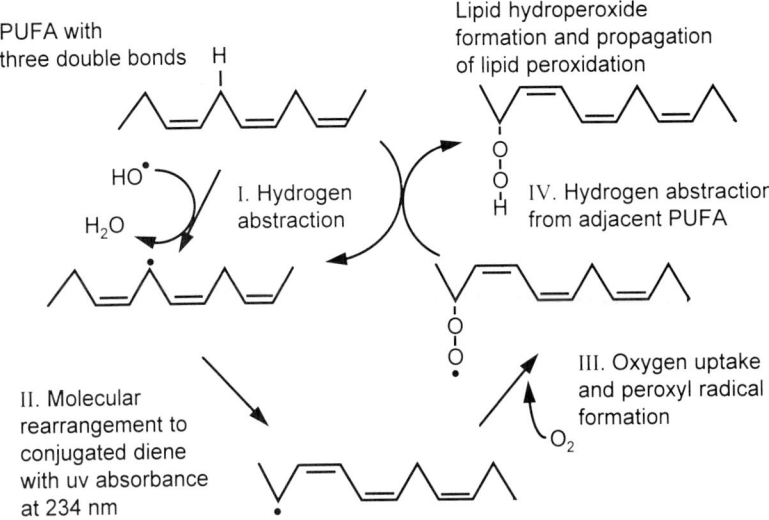

FIG. 1. Schematic representation of the autooxidation of polyunsaturated fatty acids (PUFAs). A PUFA with three double bonds undergoes hydroxyl radical–mediated hydrogen abstraction from a doubly allylic position (I), leading to molecular rearrangement and the formation of a conjugated diene (II). Conjugated dienes spontaneously combine with O_2 to form lipid peroxyl radicals (III) that abstract a hydrogen from an adjacent PUFA (IV), thus propagating the chain reaction of PUFA autooxidation. (Adapted from Keaney and Frei, 1994.)

and macrophages (Parthasarathy et al., 1986; Jialal and Grundy, 1991; Kritharides et al., 1995), are capable of oxidizing LDL by a mechanism dependent on redox-active transition metal ions, such as copper or iron. LDL can be oxidatively modified by Cu^{2+} alone (i.e., in the absence of cells) to a form that is chemically and biologically indistinguishable from cell-modified LDL (Steinbrecher et al., 1984; Esterbauer et al., 1992). In order for Cu^{2+} to initiate LDL oxidation, both Cu^{2+} binding to and reduction by LDL are required (Kuzuya et al., 1992; Lynch and Frei, 1995). Interestingly, LDL cannot reduce Fe^{3+}, and an external reductant such as superoxide radicals or thiols is required for initiation of iron-mediated LDL oxidation (Lynch and Frei, 1993, 1995). These reductants may be provided by vascular cells. For example, endothelial cells and macrophages release thiols, which, in conjunction with media-contained trace amounts of iron, can cause LDL oxidation (Heinecke et al., 1987; Sparrow and Olszewski, 1993). Endothelial cells also release superoxide radicals, which may contribute to extracellular metal ion reduction (Steinbrecher, 1988). Similarly, smooth muscle cells generate superoxide radicals and oxidize LDL by a superoxide- and metal ion-dependent mechanism (Heinecke et al., 1986).

The requirement for redox-active transition metal ions and transition metal ion reduction in cell-mediated and cell-free LDL oxidation can be readily explained by the role of these metal ions in lipid peroxidation. In LDL free of preformed lipid hydroperoxides, complexes of oxygen with reduced and oxidized metal ions (e.g. Fe^{2+}/Fe^{3+}-O_2) or ferryl ions ($FeOH^{3+}$ or FeO^{2+}) or perferryl ions (Fe^{2+}-$O_2 \leftrightarrow Fe^{3+}$-$O_2\cdot^-$), may be capable of initiating lipid peroxidation by hydrogen abstraction from a PUFA (see Fig. 1). In this regard, an Fe^{2+}:Fe^{3+} ratio of 1:1 appears to be the most potent (Minotti and Aust, 1987). In LDL containing trace amounts of preformed lipid hydroperoxides (LOOH), reduced metal ions (Me^{n+}; e.g., Cu^+ or Fe^{2+}) can catalyze the decomposition of these preformed lipid hydroperoxides to alkoxyl radicals (LO·), which subsequently reinitiate lipid peroxidation:

$$Me^{(n+1)+} \xrightarrow{\text{reductant}} Me^{n+} \qquad (1)$$

$$LOOH + Me^{n+} \longrightarrow LO\cdot + OH^- + Me^{(n+1)+} \qquad (2)$$

$$LO\cdot + LH \longrightarrow LOH + L\cdot \qquad (3)$$

$$L\cdot + O_2 \longrightarrow LOO\cdot \qquad (4)$$

$$LOO\cdot + LH \longrightarrow LOOH + L\cdot \qquad (5)$$

where LOH denotes a lipid hydroxide, LH a lipid molecule with a PUFA side chain containing a doubly allylic hydrogen, L· the corresponding carbon-centered PUFA radical, and LOO· an oxygen-centered lipid peroxyl radical. Reactions 1–3 are the reinitiation steps of lipid peroxidation, while Reactions 4 and 5 together make up the radical chain reaction of lipid peroxidation (see also Fig. 1). Since preformed lipid hydroperoxides are often present in isolated LDL, reaction sequence 1–5 is the most common mechanism of initiation of lipid peroxidation in LDL *in vitro* (see earlier).

As the availability of redox-active transition metal ions in the arterial wall is uncertain, some investigators have attempted to identify metal ion-independent mechanisms of LDL oxidation. These mechanisms include the myeloperoxidase-derived oxidants hypochlorite (Hazell and Stocker, 1993) and tyrosyl radicals (Savenkova et al., 1994), as well as NO-derived peroxynitrite (Moore et al., 1995). In addition, numerous *in vitro* studies have used the model compound 2,2′-azobis(2-amidinopropane) hydrochloride (AAPH), which generates aqueous radicals at a constant rate and initiates lipid peroxidation in LDL.

Initiation or reinitiation of lipid peroxidation is only the first step in the oxidative modification of LDL. Under some *in vitro* conditions (but not all; see Section IV,B), lipid peroxidation in LDL is initially suppressed by the endogenous antioxidants. Therefore, this initial phase of LDL oxidation has been termed the "lag phase" and is characterized by a relatively low rate of lipid peroxidation and the consumption of the LDL-associated antioxidants. The lag phase is followed by the propagation phase, during which lipid peroxidation occurs at a relatively high rate unimpeded by antioxidants (see Reactions 4 and 5 above). During the third stage of lipid peroxidation, lipid hydroperoxides decompose to form aldehydic compounds such as 4-hydroxynonenal, malondialdehyde, hexanal, and 2,4-heptadienal (Esterbauer *et al.,* 1987). Many of these aldehydic products are highly reactive and can react with the ϵ-amino groups of lysine residues in apo B, forming Schiff's bases (Haberland *et al.,* 1994; Steinbrecher, 1987). Since lysine residues are positively charged at physiological pH, the formation of Schiff's bases during the modification of LDL results in an increased net-negative charge of apo B, which can be measured as increased anodic electrophoretic mobility of LDL upon agarose gel electrophoresis. As a consequence, modified apo B is no longer recognized by the apo B/E LDL receptor (Brown and Goldstein, 1986). Instead, oxLDL containing modified apo B is internalized by macrophages via the scavenger receptor pathway (Goldstein *et al.,* 1979; Sparrow *et al.,* 1989; Freeman *et al.,* 1991).

III. Antioxidant Protection of LDL by Vitamin C

Plasma and interstitial fluid, which comprise the transport media of LDL in the human body, are endowed with a number of antioxidant species, including small molecule antioxidants and metal-binding proteins (Stocker and Frei, 1991; Frei, 1995; Dabbagh and Frei, 1995). Among these extracellular antioxidants, vitamin C (ascorbic acid) is particularly important (Frei *et al.,* 1988, 1989). Physiological ascorbic acid concentrations in human plasma and interstitial fluid range from about 30 to 100 μM (Stocker and Frei, 1991; Dabbagh and Frei, 1995). Plasma exposed to a constant flux of aqueous or lipid-soluble peroxyl radicals (Frei *et al.,* 1988, 1989, 1990; Lynch *et al.,* 1994; Neuzil and Stocker, 1994), activated neutrophils (Frei *et al.,* 1988), cigarette smoke (Frei *et al.,* 1991), or superoxide radicals and hydrogen peroxide (Frei *et al.,* 1992) is effectively protected against detectable lipid peroxidation by endogenous ascorbic acid. However, once ascorbic acid has been depleted, detectable amounts of lipid hydroperoxides are formed, despite the continued presence of other antioxidants in plasma, including α-tocopherol and β-carotene (Frei et al., 1988, 1989; Neuzil and Stocker, 1994).

Numerous *in vitro* studies have investigated the antioxidant protection of isolated human LDL by vitamin C. These studies have used many of the previously described oxidizing conditions (Section II), including cultured vascular cells, neutrophils, myeloperoxidase-derived oxidants, aqueous peroxyl radicals, and transition metal ions. Most remarkably, physiological concentrations of ascorbic acid effectively inhibit LDL oxidation under all these conditions (Table III), the only exception being activated neutrophils in the presence of ferritin (Abdalla et al., 1992). As shown in Table III, 15 to 130 μM ascorbic acid strongly suppress or abolish oxidative modification of LDL by endothelial cells (Steinbrecher, 1988; A. Martin and B. Frei, unpublished), macrophages (Jialal and Grundy, 1991; Stait and Leake, 1994), activated neutrophils (Stocker et al., 1991), tyrosyl radicals (Savenkova et al., 1994), heme and hydrogen peroxide (Balla et al., 1991; Retsky and Frei, 1995), aqueous peroxyl radicals (Retsky et al., 1993; Ma et al., 1994), Fe^{3+} in Ham's F-10 medium (A. Martin and B. Frei, unpublished), and Cu^{2+} (Jialal et al., 1990; Jialal and Grundy, 1991; Retsky et al., 1993; Hatta and Frei, 1995).

A. MOLECULAR MECHANISMS

Several mechanisms are responsible for the strong protective effects of ascorbic acid against LDL oxidation (Table IV):

1. *Scavenging of Radicals and Oxidants in the Aqueous Phase*

Ascorbic acid effectively traps free radicals and reactive oxygen species, including hypochlorite (Halliwell et al., 1987), tyrosyl radicals (Hunter et al., 1989), aqueous peroxyl radicals (Frei et al., 1988; Frei, 1995), and superoxide radicals (Nishikimi, 1985; Bendich et al., 1986). Direct scavenging of these chemical species may prevent them from attacking the lipoprotein particle and causing its oxidation. Such a mechanism likely explains the protective effects of ascorbic acid against activated neutrophils, which release copious amounts of hypochlorite capable of modifying LDL (Stocker et al., 1991; Hazell and Stocker, 1993; Hazell et al., 1994). Similarly, direct scavenging of aqueous peroxyl radicals or tyrosyl radicals by ascorbic acid may prevent these radicals from initiating lipid peroxidation in LDL (Retsky et al., 1993; Savenkova et al., 1994). Furthermore, superoxide radicals generated by endothelial cells may play a role in LDL oxidation mediated by

TABLE III
Inhibition of LDL Oxidation by Ascorbic Acid under Various Oxidizing Conditions *in Vitro*[a]

LDL oxidation system	Effect observed[b]	Vitamin C concentration (μM)	Inhibition by vitamin C (%)	References
Endothelial cells	Increased electrophoretic mobility of LDL	50	89	Steinbrecher (1988)
Macrophages	Degradation of LDL by macrophages	50	93	Jialal and Grundy (1991), Stait and Leake (1994)
	Lipid peroxidation in LDL	50	95	
	Increased electrophoretic mobility of LDL	50	93	
	Degradation of LDL by macrophages	40, 80	100	
Activated neutrophils	Lipid peroxidation in LDL	50	95	Stocker et al. (1991)
Activated neutrophils and ferritin	Lipid peroxidation in LDL	100	10	Abdalla et al. (1992)
Tyrosyl radicals	Lipid peroxidation in LDL	≥50	100	Savenkova et al. (1994)
Hemin/H_2O_2	Lag phase of lipid peroxidation in LDL	52	Extended 4-fold	Balla et al. (1991), Retsky and Frei (1995)
	Increased electrophoretic mobility of LDL	15, 60	82, 93	
Cu^{2+}	Lag phase of lipid peroxidation in LDL	30	Extended >7-fold	Jialal et al. (1990), Jialal and Grundy (1991), Retsky et al. (1993), Hatta and Frei (1995)
	Lipid peroxidation in LDL	60	100	
	Increased electrophoretic mobility of LDL	≥40	95	
	Increased electrophoretic mobility of LDL[c]	30	83	
	Degradation of LDL by macrophages	40, 60	100	
Fe^{3+}/Ham's F-10	Increased electrophoretic mobility of LDL	130	100	A. Martin and B. Frei (unpubl.)
AAPH[d]	Lag phase of lipid peroxidation in LDL	10, 50	Extended 50, 75%	Retsky et al. (1993), Ma et al. (1994)
	Lipid peroxidation in LDL	50	73	

[a] Adapted from Lynch et al. (1995).
[b] Increased electrophoretic mobility of LDL and uptake of LDL by macrophages was measured after 24 h of incubation under the indicated oxidizing conditions; lipid peroxidation refers to the initial rate of lipid hydroperoxide formation or to the amounts of thiobarbituric acid-reactive substances (TBARS) formed after 24 h of incubation.
[c] LDL was incubated at a low oxygen partial pressure (2% O_2).
[d] AAPH, 2,2'-azobis(2-amidinopropane) hydrochloride.

TABLE IV
MOLECULAR MECHANISMS OF THE ANTIOXIDANT PROTECTION OF LDL BY ASCORBIC ACID

1. Scavenging of free radicals (aqueous peroxyl, tyrosyl, and superoxide radicals) and oxidants (hypochlorite, peroxynitrite) in the aqueous milieu, preventing these radicals and oxidants from initiating lipid peroxidation in LDL and/or reducing metal ions.
2. Regeneration of LDL-associated α-tocopherol, inhibiting lipid peroxidation in LDL and preventing prooxidant activity (chain-transfer reaction) of α-tocopherol.
3. Modification of histidine residues of apo B to 2-oxo-histidine, resulting in release of bound Cu^{2+}.
4. Destruction of preformed lipid hydroperoxides in LDL in the presence of Cu^{2+}, preventing reinitiation as well as propagation of lipid peroxidation.

these cells, presumably via reduction of metal ions contained in culture media (see Section II). Therefore, scavenging of these cell-generated superoxide radicals by ascorbic acid may explain, at least in part, the observed protective effect (Steinbrecher, 1988; A. Martin and B. Frei, unpublished).

2. *Regeneration of α-Tocopherol*

Ascorbic acid may also inhibit LDL oxidation by regenerating LDL-associated α-tocopherol from the α-tocopheroxyl radical at the water–lipid interface (Sato et al., 1990; Kagen et al., 1992). For example, when radicals are generated within the LDL particle, ascorbic acid cannot directly scavenge these radicals due to different solubility and thus inhibits LDL oxidation via regeneration of LDL-associated α-tocopherol (Sato et al., 1990). However, under most other conditions, regeneration of α-tocopherol appears not to be the primary mechanism by which ascorbic acid prevents LDL oxidation (Retsky and Frei, 1995). As discussed earlier, the protective effect of ascorbic acid is more readily explained by direct scavenging of reactive species in the aqueous milieu before they can attack LDL. The contention that regeneration of α-tocopherol by ascorbic acid is not the primary mechanism of vitamin C's action is also supported by the observation that dehydroascorbic acid, the two-electron oxidation product of ascorbic acid, can preserve α-tocopherol in LDL incubated with Cu^{2+} (Retsky and Frei, 1995). Dehydroascorbic acid, has no direct reducing capability and cannot reduce α-tocopheroxyl radicals in LDL to α-tocopherol (R. Stocker and P. K. Witting, personal communication). Therefore, dehydroascorbic acid, like ascorbic acid, prevents α-tocopherol oxidation in the first place; that is, it spares, rather than regenerates, α-tocopherol (Retsky and Frei, 1995).

3. *Inhibition of Metal Ion-Binding to LDL*

Of particular interest is the observation that ascorbic acid prevents metal ion-dependent LDL oxidation, either by vascular cells in culture or in cell-free systems, using Cu^{2+}, Fe^{3+}, or hemin (Table III), because ascorbic acid is considered a pro- rather than an antioxidant in the presence of transition metal ions (Samuni *et al.*, 1983). The prooxidant effect of ascorbic acid is explained by its capability to reduce transition metal ions (see above, Reaction 1), which can lead to reinitiation of lipid peroxidation via breakdown of preformed lipid hydroperoxides (Reactions 2–5). In the absence of preformed lipid hydroperoxides, metal ion reduction, together with generation of hydrogen peroxide (H_2O_2) by ascorbic acid autooxidation, may result in production of hydroxyl radicals (HO·) via the Fenton reaction (analogous to Reaction 3) (Samuni *et al.*, 1983; Bendich *et al.*, 1986):

$$H_2O_2 + Me^{n+} \rightarrow HO\cdot + HO^- + Me^{(n+1)+} \tag{6}$$

Hydroxyl radicals are extremely reactive species that can initiate lipid peroxidation in LDL (Bedwell *et al.*, 1989). Based on these considerations, ascorbic acid should promote, rather than prevent, metal ion-dependent LDL oxidation.

In a partial explanation for the paradoxical protective effects of vitamin C against metal ion-dependent LDL oxidation, dehydroascorbic acid was found to effectively prevent LDL oxidation by hemin or Cu^{2+} (Retsky *et al.*, 1993; Retsky and Frei, 1995) (Fig. 1). Both ascorbic acid and dehydroascorbic acid prevented the various stages of LDL oxidative modification: consumption of endogenous antioxidants (except for ubiquinol-10), initiation of lipid peroxidation, apo B modification, and degradation of LDL by macrophages. In addition, in LDL incubated with Cu^{2+} and ascorbic acid or dehydroascorbic acid, histidine residues of apo B were oxidized to 2-oxo-histidine, and 2-oxo-histidine formation was associated with loss of bound Cu^{2+} from LDL (K. L. Retsky, K. Chen, and B. Frei, unpublished). These data are in agreement with previous findings showing that 2-oxo-histidine, in contrast to histidine, cannot bind Cu^{2+} (Uchida and Kawakishi, 1990). As Cu^{2+} binding to apo B is a prerequisite for LDL oxidation (see Section II), this limited, site-specific oxidative "damage" to the histidine residues of apo B resulting in loss of bound Cu^{2+} could explain the paradoxical protective effects of vitamin C against Cu^{2+}-mediated LDL oxidation.

4. *Destruction of Preformed Lipid Hydroperoxides*

In addition to the previous mechanisms, ascorbic acid also prevents propagation of lipid peroxidation in LDL by rapid and complete destruction of preformed lipid hydroperoxides. This effect of ascorbic acid was dependent on redox-active copper and was observed in LDL during the early stages of the propagation phase of lipid peroxidation (Retsky and Frei, 1995). In contrast, in extensively oxidized LDL, ascorbic acid addition may have a prooxidant effect (Stait and Leake, 1994) due to metal ion reduction and stimulation of lipid peroxidation via alkoxyl radical formation (see Reactions 1–5 above). Prevention of Cu^{2+} binding to LDL may no longer be a critical factor in these advanced stages of LDL oxidation, where lipid peroxidation can occur by a self-propagating mechanism (Reactions 4 and 5), independent of initiating events.

5. Summary

The data discussed here demonstrate that vitamin C strongly protects LDL against oxidation by several distinct mechanisms (Table IV). These mechanisms may act in concert to prevent modification of LDL *in vivo*, irrespective of whether the latter occurs by metal ion-dependent or -independent oxidation. Vitamin C, therefore, by inhibiting LDL oxidative modification, may inhibit initiation and progression of atherosclerotic lesion development and lower the risk of myocardial infarction and ischemic stroke.

IV. LDL Oxidation and Vitamin E

Human LDL is protected not only by the antioxidant defenses present in plasma and interstitial fluids, but also by lipid-soluble antioxidants associated with the LDL particle itself (Table II). As mentioned earlier, the most abundant antioxidant in human LDL is α-tocopherol (vitamin E), with approximately five to nine molecules per LDL particle (Esterbauer *et al.*, 1989, 1992; Frei and Gaziano, 1993). α-Tocopherol is known to act as a chain-breaking antioxidant by scavenging lipid peroxyl radicals:

$$\alpha\text{-tocH} + \text{LOO·} \rightarrow \alpha\text{-toc·} + \text{LOOH} \tag{7}$$

$$\alpha\text{-toc·} + \text{LOO·} \rightarrow \alpha\text{-toc-OOL} \tag{8}$$

In Reaction 7, α-tocopherol donates its phenolic hydrogen to a lipid peroxyl radical that otherwise would propagate the radical chain reaction of lipid peroxidation (see Reaction 5). The α-tocopheroxyl radical formed in Reaction 7 may react with another lipid peroxyl radical, forming a covalent adduct (Reaction 8). Thus, each molecule of α-tocopherol can scavenge two peroxyl radicals, thereby breaking the chain propagation reaction of lipid peroxidation.

During the lag phase of lipid peroxidation in LDL, the intraparticle antioxidants are consumed in a distinct sequence. The first antioxidant consumed is ubiquinol-10, followed by α- and γ-tocopherol, lycopene, and β-carotene (Esterbauer *et al.*, 1989; Stocker *et al.*, 1991; Lynch *et al.*, 1994). The depletion of the LDL-associated antioxidants is followed by the propagation phase of lipid peroxidation and, eventually, modification of apo B (Lynch *et al.*, 1994) and recognition of LDL by the macrophage scavenger receptors (see Section II).

A. ANTIOXIDANT PROTECTION OF LDL

The observation of a substantially increased rate of lipid peroxidation in LDL after consumption of intraparticle antioxidants suggested that the resistance of LDL to oxidative modification is related to its antioxidant content (Esterbauer *et al.*, 1989, 1991; Dieber-Rotheneder *et al.*, 1991). Interestingly, however, numerous *in vitro* studies have shown that the lipid peroxidation lag phase in LDL is not related to the α-tocopherol content of LDL, unless α-tocopherol levels have been raised substantially by supplementation *in vivo* (see Section V,B) or *in vitro* (see later). For example, Esterbauer *et al.* (1989, 1992) reported a nonsignificant correlation coefficient of $R^2 = 0.043$ for the resistance to Cu^{2+}-induced oxidation and α-tocopherol content of 78 LDL samples from non-vitamin E-supplemented donors, and Frei and Gaziano (1993) observed a similar nonsignificant R^2 value of 0.063 for 61 LDL samples. Very similar results were obtained by numerous other investigators (Babiy *et al.*, 1990; Jessup *et al.*, 1990; Cominacini *et al.*, 1991; Smith *et al.*, 1993; Miller *et al.*, 1995). Frei and Gaziano (1993) estimated that only about 20% of the lipid peroxidation lag phase in Cu^{2+}-incubated LDL can be explained by the α-tocopherol content of LDL, and Esterbauer *et al.* (1992) came to basically the same conclusion. LDL resistance to Cu^{2+}-induced oxidation also did not correlate with the total LDL antioxidant content (Esterbauer *et al.*, 1989, 1992). Therefore, other parameters seem to significantly contribute to the susceptibility of LDL to oxidation, such as preformed lipid hydroperoxides (Frei and Gaziano, 1993), PUFA content (Reaven *et al.*, 1991), and metal-binding capacity (Gieseg and Esterbauer, 1994).

An alternative approach that has been used to investigate the role of α-tocopherol in inhibiting LDL oxidation is direct addition of α-tocopherol to LDL oxidation systems *in vitro*. Using this method, it was found that α-tocopherol limits the oxidative modification of LDL by cultured vascular cells (Morel *et al.*, 1984; Steinbrecher *et al.*, 1984; van Hinsbergh *et al.*, 1986; Leake and Rankin, 1990) and Cu^{2+} (Esterbauer *et al.*, 1989, 1991; Jialal *et al.*, 1990; Jialal and Grundy, 1991). Most of these studies, however, used α-tocopherol concentrations greatly exceeding those found *in vivo*. For example, α-tocopherol concentrations in human plasma and interstitial fluid are about 25 and 5 µ*M*, respectively (Dabbagh and Frei, 1995), yet in studies of the effects of added α-tocopherol on endothelial cell-mediated LDL oxidation, α-tocopherol concentrations of 100 µ*M* (Steinbrecher *et al.*, 1984) and 230 µ*M* (van Hinsbergh *et al.*, 1986) were used. When physiological or near-physiological concentrations of α-tocopherol were added, the effects on LDL oxidation were less pronounced: Esterbauer *et al.* (1989) observed a 1.6-fold extension of the lipid peroxidation lag phase in Cu^{2+}-exposed LDL following addition of 4.6 µ*M* α-tocopherol, and Jialal *et al.* (1990) found an approximately 50% reduction in electrophoretic mobility and macrophage degradation of LDL incubated for 24 h with Cu^{2+} in the presence of 40 µ*M* α-tocopherol, compared to 95–100% inhibition by 40 µ*M* ascorbic acid (see Table III).

The requirement for nonphysiological concentrations of α-tocopherol to substantially inhibit LDL oxidation *in vitro* may reflect poor α-tocopherol incorporation into LDL under these conditions. Esterbauer *et al.* (1991) have shown that adding α-tocopherol in ethanol to LDL in aqueous solution does not result in effective incorporation of α-tocopherol into LDL. In order to achieve significant increases in LDL α-tocopherol content *in vitro*, plasma must be incubated with excessive amounts of α-tocopherol prior to LDL isolation. Using this method, 1–10% of the added α-tocopherol is incorporated into LDL, providing for a 2- to 10-fold enhancement in the LDL α-tocopherol content (Esterbauer *et al.*, 1991; Hatta and Frei, 1995). LDL treated in this manner demonstrated enhanced resistance to Cu^{2+}-induced oxidation, as assessed by the lag phase of lipid peroxidation (Esterbauer *et al.*, 1991; Hatta and Frei, 1995). In contrast to unsupplemented LDL (see earlier), in α-tocopherol-supplemented LDL the lag phase was highly significantly correlated with the α-tocopherol content relative to protein ($R^2 = 0.96$; $p<0.001$) (Esterbauer *et al.*, 1991). Despite this tight correlation, the absolute effects of α-tocopherol enrichment on LDL oxidation were modest: a 5-fold increase in LDL α-tocopherol content was associated with a 1.4-fold extension of the lipid peroxidation lag phase but no reduction in electrophoretic mobility following 24 h of incubation with Cu^{2+} (Hatta and Frei, 1995). Similarly, when LDL containing 5-fold-increased α-tocopherol levels was incubated with Cu^{2+} under an oxygen partial pressure (corresponding to 2% O_2) close to physiological pressures in the arterial wall (2.5–10% O_2) (Hajjar et al., 1988; Crawford and Blankenhorn, 1991), rather than ambient oxygen partial pressure (about 20% O_2), there was a 50% reduction in the initial rate of lipid peroxidation but again no inhibitory effect on increased electrophoretic mobility (Hatta and Frei, 1995). Because increased electrophoretic mobility of LDL correlates well with macrophage degradation, and thus increased atherogenicity of LDL (Steinberg et al., 1989; Keaney and Frei, 1994), these data suggest that α-tocopherol may have little effect on atherogenic modification of LDL in vivo.

B. In Vitro Stimulation of LDL Oxidation

Evidence shows that, under certain in vitro conditions, α-tocopherol acts as a prooxidant and stimulates, rather than inhibits, lipid peroxidation in LDL (Bowry et al., 1992; Bowry and Stocker, 1993; Ingold et al., 1993). In these reactions, α-tocopherol acts as a chain-transfer agent rather than as a chain-breaking antioxidant; that is, the α-tocopheroxyl radical, formed, for example, in Reaction 7 or by reaction with Cu^{2+} (Yoshida et al., 1994; Lynch and Frei, 1995) (Fig. 2), does not trap a lipid peroxyl radical (Reaction 8), but instead abstracts a doubly allylic hydrogen from a PUFA side chain:

$$\alpha\text{-toc}\cdot + LH \rightarrow \alpha\text{-tocH} + L\cdot \qquad (9)$$

The carbon-centered lipid radical (L·) formed in Reaction 9 then initiates the lipid peroxidation chain reaction, which leads to formation of another lipid peroxyl radical. The latter may react anew with an α-tocopherol molecule to form an α-tocopheroxyl radical, completing the cycle (Fig. 2). This reaction sequence has been termed "tocopherol-mediated peroxidation" (Bowry and Stocker, 1993). Reaction 9 is a chain-transfer reaction, because the α-tocopheroxyl radical passes the radical character on to a lipid. In effect, Reaction 9 can replace Reaction I shown in Fig. 1. Therefore, lipid peroxidation in LDL in vitro may be initiated by three different mechanisms: (i) de novo initiation (Fig. 1), (ii) reinitiation by breakdown of preformed lipid hydroperoxides (Reactions 1–3); and (iii) α-tocopheroxyl radical-mediated initiation (Fig. 2).

There is now compelling and conclusive evidence for tocopherol-mediated peroxidation in LDL incubated under various in vitro conditions. For example, when LDL is incubated in Ham's F-10 medium (containing trace amounts of iron and copper), the rate of lipid peroxidation is maximal during α-tocopherol consumption but declines

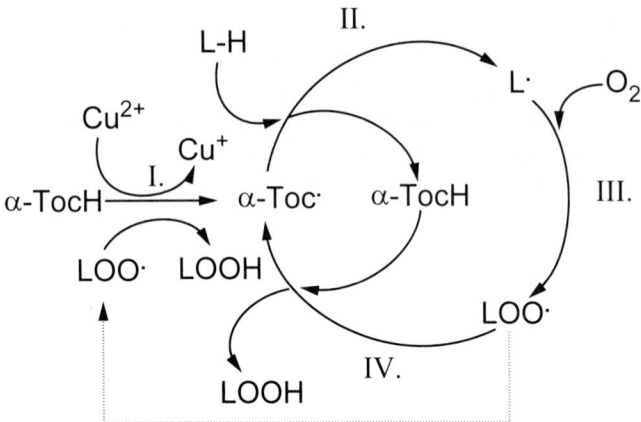

FIG. 2. Tocopherol-mediated peroxidation. In Reaction I, α-tocopherol (α-TocH) donates its phenolic hydrogen to a lipid peroxyl radical (LOO·), forming a lipid hydroperoxide (LOOH) and an α-tocopheroxyl radical (α-Toc·). Alternatively, the α-tocopheroxyl radical may be formed by reaction of α-tocopherol with Cu^{2+}. In Reactions II–IV, the α-tocopheroxyl radical acts as a chain-transfer agent by abstracting a hydrogen from a polyunsaturated fatty acid (L-H), leading to formation of a carbon-centered lipid radical (L·) and subsequently a lipid peroxyl radical (LOO·). The latter can be scavenged by another α-tocopherol molecule, leading to formation of a lipid hydroperoxide and another α-tocopheroxyl radical, thus completing the cycle of tocopherol-mediated peroxidation. Increasing the LDL content of α-tocopherol serves to facilitate the formation of α-tocopheroxyl radicals and enhance tocopherol-mediated peroxidation. This scheme is dependent upon a lack of suitable coantioxidants (i.e., ascorbic acid or ubiquinol-10) that would export the radical character from the LDL particle. (Adapted from Keaney and Frei, 1994.)

thereafter (Bowry and Stocker, 1993). Similar results were obtained when LDL depleted of endogenous ubiquinol-10 was exposed to aqueous radicals generated by AAPH at a low rate (i.e., <250 nM/min) (Bowry et al., 1992; Ingold et al., 1993). Furthermore, *in vivo* or *in vitro* enrichment of LDL with α-tocopherol was associated with accelerated LDL oxidation by both aqueous and lipophilic peroxyl radicals, again at low radical fluxes (Bowry et al., 1992; Ingold et al., 1993). In agreement with these observations, Frei and Gaziano (1993) reported that the vitamin E content of LDL from nonsupplemented subjects is positively associated with the susceptibility of LDL to AAPH-induced oxidation. Finally, LDL from patients with familial isolated vitamin E deficiency, which contains very little to no detectable α-tocopherol, is much more resistant to *in vitro* oxidation induced by various agents than control LDL containing normal levels of α-tocopherol (R. Stocker, personal communication). The same effect is seen when LDL is depleted of α-tocopherol *in vitro* prior to incubation of LDL with AAPH or Cu^{2+}, or in Ham's F-10 medium (S. M. Lynch and B. Frei, unpublished).

The following specific example illustrates the prooxidant activity of LDL-associated α-tocopherol *in vitro*. Under experimental conditions using 1 mg/ml of LDL protein incubated with 1 mM AAPH at 37°C (i.e., when radicals are generated at a rate of about 60 nM/min), LDL encounters only one aqueous peroxyl radical every 30 min (Ingold *et al.*, 1993). Not every encounter of a water-soluble peroxyl radical with an LDL particle will result in transfer of the radical character from the water milieu into the lipoprotein. Because their doubly allylic hydrogens are buried within the LDL particle, the phospholipid molecules in the LDL surface are rather resistant to attack by water-soluble peroxyl radicals. In contrast, the phenol group of α-tocopherol is located close to the LDL surface, and is among the most reactive species in LDL facing the aqueous milieu, together with the reactive moieties of ubiquinol-10 and preformed lipid hydroperoxides (Ingold *et al.*, 1993). Therefore, in the absence of preformed lipid hydroperoxides and ubiquinol-10, α-tocopherol will facilitate the transfer of the radical character from the aqueous phase into the LDL particle. This reaction is also termed a "phase-transfer" reaction, which refers to hydrogen abstraction from the phenol group of α-tocopherol by an aqueous radical ($X\cdot_{aq}$) at the water–lipid interface, resulting in formation of an α-tocopheroxyl radical in LDL:

$$\alpha\text{-tocH}_{LDL} + X\cdot_{aq} \rightarrow \alpha\text{-toc}\cdot_{LDL} + XH_{aq} \qquad (10)$$

Thus, LDL-associated α-tocopherol can act as a prooxidant *in vitro* by two mechanisms: the chain-transfer reaction (Reaction 9; Fig. 2) and the phase-transfer reaction (Reaction 10). Once formed within the LDL particle, the mean life span of an α-tocopheroxyl radical is approximately 12.5 s (Ingold *et al.*, 1993). The relative stability of the α-tocopheroxyl radical and its isolation from other LDL particles increases the likelihood of a chain-transfer reaction with an adjacent PUFA side chain.

It is important to point out that the prooxidant chain-transfer reaction of the α-tocopheroxyl radical is much less likely to occur under the following circumstances:

1. *The rate of radical production is high,* and several radicals are present in LDL or its immediate aqueous environment. This allows the α-tocopheroxyl radical to react with another radical, resulting in chain termination and inhibition of lipid peroxidation

(Bowry et al., 1992; Bowry and Stocker, 1993; Ingold et al., 1993). Therefore, the prooxidant effect of LDL-associated α-tocopherol is only observed at low radical fluxes (see earlier).
2. *The LDL particle containing the α-tocopheroxyl radical is not isolated from other LDL particles;* that is, it can transfer between LDL particles, thereby increasing the likelihood of encountering and reacting with another (α-tocopheroxyl) radical. Therefore, LDL-associated α-tocopherol may act as a prooxidant in LDL suspended in aqueous solution but not in LDL aggregates.
3. *Co-antioxidants are present,* such as ubiquinol-10, ascorbic acid, or bilirubin (Ingold et al., 1993; Neuzil and Stocker, 1994). These co-antioxidants can export the radical character from within the LDL particle back to the aqueous milieu, thus avoiding the perils of a chain-transfer reaction by the α-tocopheroxyl radical in LDL (Ingold et al., 1993).
4. *LDL contains substantial amounts of preformed lipid hydroperoxides,* in which case lipid peroxidation is predominantly reinitiated (Iwatsuki et al., 1995). Because the preformed lipid hydroperoxides are exposed at the LDL surface, they are likely to capture radicals from the aqueous phase. This reinitiation mechanism of lipid peroxidation may override the possible prooxidant effect of α-tocopherol, allowing α-tocopherol to act as an antioxidant.

These considerations are germane to the question of whether α-tocopherol in LDL acts as a pro- or antioxidant *in vivo,* and, therefore, the question of whether vitamin E supplementation may accelerate or inhibit the atherosclerotic process. It is likely that, in the arterial wall, no preformed lipid hydroperoxides are initially present in LDL and the rate of radical production is low; these conditions increase the likelihood of a prooxidant activity of α-tocopherol. Conversely, LDL may form aggregates, and coantioxidants such as ubiquinol-10 and ascorbic acid are likely to be present in the arterial wall, which increases the likelihood of α-tocopherol acting as an antioxidant. Epidemiological studies tend to indicate a protective effect of supplemental α-tocopherol against coronary artery disease (Gaziano et al., 1994), but whether this is due to decreased LDL oxidation or some other effect of α-tocopherol remains to be seen. In addition, studies indicate that human atherosclerotic plaques contain large amounts of lipid peroxidation products coexisting with large amounts of α-tocopherol, suggesting that the latter did not effectively prevent lipid peroxidation (R. Stocker, personal communication). Finally, results from animal studies of the effects of α-tocopherol supplementation on atherosclerotic lesion development are mixed and inconsistent, with some studies showing proatherogenic effects of α-tocopherol (Lynch and Frei, 1994).

V. Trials of *In Vivo* Antioxidant Supplementation

The studies discussed previously showing that vitamin C and vitamin E supplemented *in vitro* can limit LDL oxidation (Sections III and IV,A) have stimulated a number of small-scale clinical trials examining the effect of dietary vitamin C and vitamin E supplementation on the resistance of plasma-derived LDL to oxidation *in vitro*. In light of the evidence discussed earlier for chain-transfer and phase-transfer activity of LDL-associated α-tocopherol *in vitro*, it must be realized that those studies in Section IV,A and Section V,B that found a protective effect of α-tocopherol must have used *in vitro* conditions favoring an antioxidant effect of α-tocopherol. This can be readily explained by the facts that isolated LDL often contains preformed lipid hydroperoxides and is exposed to high rates of radical production *in vitro*, conditions under which α-tocopherol is unlikely to act as a prooxidant. However, as discussed earlier, the relevance of these studies showing a protective effect of α-tocopherol is unclear, as the conditions of LDL oxidation in the arterial wall are unknown. The small-scale clinical trials of the effects of *in vivo* vitamin C and E supplementation on LDL oxidation are discussed in this section.

A. Vitamin C

The effect of oral vitamin C supplementation on LDL antioxidant protection has not been studied in any great detail. The main reason for this is that ascorbic acid is removed from LDL during isolation from plasma, and thus *in vivo* vitamin C supplementation is bound to show no effect on oxidation of isolated LDL *in vitro*! It comes as no surprise, therefore, that *in vivo* vitamin C supplementation in humans is without effect (Belcher *et al.*, 1993; Reaven *et al.*, 1993) or only moderately increases (Harats *et al.*, 1990; Rifici and Khachadurian, 1993) the resistance of plasma-derived LDL to Cu^{2+}-induced or cell-mediated oxidation *in vitro*. The moderately increased oxidative resistance of LDL following vitamin C supplementation observed in some studies (Harats *et al.*, 1990; Rifici and Khachadurian, 1993) has been attributed to the preservation of LDL-associated vitamin E by vitamin C. However, this notion could not be substantiated by measurement of the actual plasma levels of these vitamins (Harats *et al.*, 1990).

These negative results should not be interpreted as indicating that vitamin C has no protective effect against LDL oxidation *in vivo*, where LDL and ascorbic acid coexist in plasma and other extracellular fluids (Dabbagh and Frei, 1995). Epidemiological studies of the role of dietary vitamin C in reducing the risk of coronary artery disease are inconclusive, whereas several studies found a significantly and substantially reduced risk of ischemic stroke with increased vitamin C consumption (Gaziano *et al.*, 1994). Data from animal studies suggest that vitamin C deficiency is associated with accelerated atherosclerosis (Lynch *et al.*, 1995). Whether the latter finding is due to decreased antioxidant protection of LDL in the absence of sufficient amounts of vitamin C or some other, metabolic effect of vitamin C is unclear. Further studies are needed to answer the question of whether vitamin C can inhibit LDL oxidation and atherosclerosis in humans.

B. Vitamin E

Several small-scale clinical trials have examined the effect of dietary α-tocopherol supplementation on LDL antioxidant protection *in vitro*. Some of these trials have used antioxidant combinations (Princen *et al.*, 1992; Reaven *et al.*, 1993; Jialal and Grundy, 1993; Abbey *et al.*, 1993), while others have dealt with α-tocopherol exclusively (Dieber-Rotheneder *et al.*, 1991; Jialal and Grundy, 1992; Reaven and Witztum, 1993; Jialal *et al.*, 1995; Princen *et al.*, 1995). Those studies using antioxidant combinations, usually α-tocopherol, vitamin C, and β-carotene, demonstrated no effect of vitamin C (see earlier) and β-carotene on LDL oxidation. Any beneficial effect, therefore, was explained by α-tocopherol alone.

Dieber-Rotheneder and colleagues (1991) performed the first study of the effects of dietary α-tocopherol supplementation in humans on the resistance of plasma-derived LDL to *in vitro* Cu^{2+}-induced oxidation. In this study, eight subjects were divided into four groups of two that received 150, 225, 800, of 1200 IU/day of α-tocopherol (as *RRR*-α-tocopherol) over a 3-week period. These doses far exceed the U.S. recommended dietary allowance of 12 IU in women and 15 IU in men (National Research Council, 1989). At the end of the supplementation period, plasma from these subjects demonstrated an increase in α-tocopherol content of 68, 91, 155, and 196%, respectively, which correlated with increases in LDL α-tocopherol content by 54, 106, 53, and 141%, respectively. LDL isolated from the subjects in all four supplementation groups and incubated *in vitro* with Cu^{2+} exhibited increased lipid peroxidation lag phases compared to LDL isolated from the same subjects before supplementation.

In a similar study, Jialal and Grundy (1992) treated 12 subjects with 800 IU/day of α-tocopherol over a 12-week period, and an equal number of subjects received placebo. The LDL α-tocopherol content was not examined but plasma α-tocopherol concentrations increased 5-fold in the α-tocopherol group, while no significant changes were observed in the placebo group. Compared to baseline, LDL obtained after α-tocopherol treatment exhibited a 2.2-fold increase in the lag phase and a reduced content of thiobarbituric acid-reactive substances (TBARS, an indirect measure of lipid peroxidation) in response to oxidation with Cu^{2+}. The same investigators conducted a study of combined antioxidant supplementation (800 IU/day of α-tocopherol, 1 g/day of vitamin C, and 30 mg/day of β-carotene), and found significant inhibition of Cu^{2+}-induced TBARS formation and a 2.4-fold prolongation of the lag phase in LDL obtained from the 12 subjects after as compared to prior to the 12-week supplementation period (Jialal and Grundy, 1993). As in the previous study, the placebo group ($N = 12$) did not show any significant changes in LDL resistance to *in vitro* oxidation. By comparing the results of this study (Jialal and Grundy, 1993) to those obtained earlier with α-tocopherol supplementation alone (Jialal and Grundy, 1992), the authors concluded that the combined protective effect of α-tocopherol, vitamin C, and β-carotene was not greater than the effect of α-tocopherol supplementation alone. However, the validity of both of these studies has been questioned (Frei and Lynch, 1994).

While the studies by Dieber-Rotheneder *et al.* (1991) and Jialal and Grundy (1992, 1993) used α-tocopherol supplementation for 3 and 12 weeks, respectively, a study by Princen and colleagues (1992) demonstrated that shorter periods of α-tocopherol supplementation also can enhance antioxidant protection of LDL. These investigators gave 1,000 IU/day of α-tocopherol to seven healthy volunteers for only 1 week. Plasma obtained from these subjects demonstrated a 3-fold increase in α-tocopherol concentration, and LDL exhibited a 2.5-fold increase in α-tocopherol content compared to LDL obtained prior to dietary supplementation. LDL derived from α-tocopherol–treated subjects and incubated with Cu^{2+} exhibited a significantly increased lag phase (1.4-fold, from 108 ± 6 to 152 ± 17 min; $p < 0.001$) and a 19% decrease in the propagation rate of lipid peroxidation.

As part of a long-term study of combined antioxidant treatment, Reaven and colleagues (1993) supplemented seven subjects with 1600 mg/day of α-tocopherol over a 5-month period. Compared to baseline values, treatment was associated with a 2.5-fold increase in LDL α-tocopherol content and enhanced resistance of LDL to oxidation by endothelial cells or Cu^{2+}. LDL from treated subjects exposed to Cu^{2+} or cells also demonstrated diminished macrophage degradation compared to LDL obtained prior to α-tocopherol supplementation. In a related study, Reaven and Witztum (1993) compared the effects of supplementation with 1600 mg/day of naturally occurring RRR-α-tocopherol to the synthetic racemic form of α-tocopherol, and equal protection of LDL against *in vitro* oxidation by endothelial cells or Cu^{2+} was found.

Abbey and colleagues (1993) examined the effect of combined supplementation with α-tocopherol, β-carotene, and vitamin C on resistance of plasma-derived LDL to *in vitro* Cu^{2+}-induced oxidation. In this study, 22 subjects (10 men and 12 women) received α-tocopherol (200 mg/day), β-carotene (18 mg/day), and vitamin C (900 mg/day) for a period of 6 months. After 3 months of supplementation, plasma levels of α-tocopherol, β-carotene, and vitamin C increased 1.6-, 5-, and 1.3-fold, respectively. LDL was more resistant to *in vitro* oxidation, as determined by the lipid peroxidation lag phase, after both 3 and 6 months of supplementation compared to baseline. In addition, there was a significant correlation between prolongation of the lag phase and LDL α-tocopherol content ($R^2 = 0.22$; $p<0.01$), but no correlation was observed with β-carotene content of LDL. Therefore, in agreement with other studies (Jialal and Grundy, 1993; Reaven *et al.*, 1993), the study by Abbey *et al.* (1993) indicated that supplementation with a combination of α-tocopherol, β-carotene, and vitamin C is associated with enhanced resistance of LDL to *in vitro* Cu^{2+}-induced oxidation due to increased α-tocopherol content of LDL.

Two recent studies have focused on the minimal doses of α-tocopherol required to significantly increase the resistance of LDL to oxidation, yielding widely differing results (Jialal *et al.*, 1995; Princen *et al.*, 1995). One previous study (Dieber-Rotheneder *et al.*, 1991) also had used a range of α-tocopherol doses (see earlier), but the number of subjects in each group ($N = 2$) was not sufficient for statistical analysis between the groups. The recent study by Jialal *et al.* (1995) used eight subjects per group and tested the effects of supplementation with 60, 200, 400, 800, and 1200 IU/day of α-tocopherol. There was a dose-dependent increase in plasma α-tocopherol levels, and a maximal 3.5-fold increase (from 22 to 78 μM) was observed in the group receiving 1200 IU of α-tocopherol per day. α-Tocopherol content of LDL also tended to be increased dose dependently, with a 2.8-fold increase (from 8 to 22 nmol/mg protein) in the group receiving 1200 IU/day. Interestingly, LDL isolated at the end of the 8-week supplementation period from the subjects receiving 60 or 200 IU/day did not show significantly increased resistance to *in vitro* Cu^{2+}-induced oxidation, in contrast to LDL from subjects given 400, 800, or 1200 IU/day. The highest dose of α-tocopherol supplementation resulted in a 1.7-fold extension of the lipid peroxidation lag phase in isolated LDL (from 81±10 to 139±35 min), as measured by TBARS formation. The authors concluded that the minimum doses of α-tocopherol needed to significantly increase the resistance of LDL to oxidation is 400 IU/day (Jialal et al., 1995).

The study by Princen et al. (1995) came to a very different conclusion. In this study, 20 subjects (10 men and 10 women) ingested consecutively 25, 50, 100, 200, 400, and 800 IU of α-tocopherol per day during six 2-week periods. At the end of each period, LDL was isolated and analyzed for α-tocopherol content and resistance to oxidation. A linear increase in LDL α-tocopherol content was observed over the entire study period up to the maximal dose, from 13 nmol/mg protein at baseline to 30 nmol/mg protein at the end of the study. Plasma α-tocopherol levels also increased steadily from 24 to 61 μM, and were well correlated with LDL α-tocopherol content ($R^2 = 0.43$; $p<0.0001$). Interestingly, after ingestion of only 25 IU/day of α-tocopherol for 2 weeks, the lipid peroxidation lag phase in LDL incubated with Cu^{2+} increased significantly by 5.3% compared to baseline, and further increased dose dependently to a maximal increase of 28% with 800 IU/day. While this study indicated that the minimal supplementary dose of α-tocopherol necessary to protect LDL against oxidation *in vitro* is 25 IU/day, the absolute benefits observed even at the highest dose of α-tocopherol were very modest.

Thus, the studies discussed here clearly demonstrate that dietary supplementation with α-tocopherol results in significant increases in plasma and LDL α-tocopherol content, as well as enhanced resistance of LDL to *in vitro* oxidation. However, the significance of these results is unclear. First, as discussed earlier, all of these studies used *in vitro* conditions of LDL oxidation that favored the antioxidant over the pro-oxidant activity of α-tocopherol, and it is currently not known which conditions best mimic those *in vivo*. While the most commonly used oxidant species in the *in vivo* supplementation studies to oxidize LDL was Cu^{2+}, the availability of redox-active copper *in vivo* is unclear. Under other, more relevant oxidizing conditions, different results may be obtained. Second, it should be noted that, even with massive doses of vitamin E (up to 1200 IU/day), only modest increases in the resistance of LDL to oxidation were observed. This is in agreement with the *in vitro* studies showing that a 5-fold increase in LDL α-tocopherol content is associated with only a 1.4-fold extension of the lipid peroxidation lag phase (Hatta and Frei, 1995). Whether such modest increases in LDL resistance to oxidation have any significant effects on the progression of atherosclerosis in humans is not known.

VI. CONCLUSIONS

According to the oxidative modification hypothesis of atherosclerosis, LDL entrapped in the subendothelial space of lesion-prone arterial sites becomes oxidized, possibly through the action of resident vascular cells. The accumulation and oxidative modification of LDL is associated with the recruitment of monocytes and macrophages that internalize oxLDL, leading to foam cell formation and development of fatty streaks. The continued presence of oxLDL in the vascular wall contributes to abnormalities in the local control of platelet adhesion, vascular tone, and the response to arterial injury. All of these features combine to produce atherosclerotic progression and a fertile environment for the development of acute vascular syndromes such as angina pectoris, myocardial infarction, and ischemic stroke resulting from plaque rupture, local thrombosis, and abnormal vascular function.

Human LDL and extracellular fluids contain a number of antioxidant defense mechanisms that may inhibit these processes, and thus may limit the development of atherosclerosis. There is a large body of evidence suggesting that LDL oxidation *in vitro* can be inhibited or prevented by vitamins C and E. Physiological concentrations of ascorbic acid very effectively prevent *in vitro* LDL oxidation under many different types of oxidizing conditions (see Tables III and IV). In contrast, there is considerable controversy as to whether α-tocopherol acts as a pro- or antioxidant in LDL. The chemistry of the chain-transfer and phase-transfer activity of α-tocopherol is well understood, and there can be no doubt that, under certain *in vitro* conditions, LDL-associated α-tocopherol promotes, rather than inhibits, LDL oxidation. Therefore, the most salient questions that must be addressed in future studies are those inquiring into the precise conditions of *in vivo* LDL oxidation: What is the rate of radical production in the arterial wall? Are coantioxidants such as ubiquinol-10 and ascorbic acid present in the arterial wall and replenished on a continuous basis, or are they depleted over time and sometimes absent? What is the role, if any, of preformed lipid hydroperoxides in initiation of LDL oxidation *in vivo*? What is the nature of the oxidizing species that initiates LDL oxidation in the arterial wall? Without this information, trials of *in vivo* α-tocopherol supplementation showing increased resistance of isolated LDL to oxidation under the specific *in vitro* conditions employed are not very informative. In addition to the previous questions, the role and significance of oxLDL itself as a causal factor in the pathogenesis of atherosclerosis must be further documented and established.

Thus, due to the lack of a better understanding of the atherosclerotic process and the role and molecular mechanism of LDL oxidation *in vivo*, the data showing significant antioxidant protection of LDL by vitamins C and E *in vitro* cannot and should not be extrapolated to the *in vivo* situation in humans. Epidemiological data and animal studies suggest that antioxidant supplementation holds promise in limiting the development of atherosclerosis, but even in these studies it is unclear whether the antioxidants act via inhibition of LDL oxidation or some other mechanism (Gaziano *et al.*, 1994; Lynch and Frei, 1994). In fact, emerging data from numerous animal studies suggest that antioxidants can act as antiatherogens without inhibiting LDL oxidation (Keaney *et al.*, 1993; Shaish *et al.*, 1995). Many other potential mechanisms of antioxidant action of vitamins C and E must be considered (Keaney and Frei, 1994). It may even be that vitamins C and E affect the atherosclerotic process through nonantioxidant mechanisms, such as changes in lipoprotein distribution, effects on thrombosis and fibrinolysis, regulation of vascular tone, and proliferation of smooth muscle cells (Keaney and Frei, 1994; Lynch *et al.*, 1995). The resolution of these issues will be pivotal for a better understanding of the pathogenesis of atherosclerosis and a critical assessment of the ability of vitamins C and E to significantly reduce morbidity and mortality from atherosclerotic vascular disease in humans.

ACKNOWLEDGMENT

The work in B. Frei's laboratory was supported by the National Institutes of Health grants HL-49954, HL-56170 and ES-06593.

REFERENCES

Abbey, M., Nestel, P. J., and Baghurst, P. A. (1993). Antioxidant vitamins and low-density-lipoprotein oxidation. *Am. J. Clin. Nutr.* **58**, 525–532.

Abdalla, D. S. P., Campa, A., and Monteiro, H. P. (1992). Low density lipoprotein oxidation by stimulated neutrophils and ferritin. *Atherosclerosis* **97**, 149–159.

Aviram, M. (1989). Modified forms of low density lipoprotein affect platelet aggregation in vitro. *Thromb. Res.* **53**, 561–567.

Babiy, A., Gebicki, J. M., and Sullivan, D. R. (1990). Vitamin E content and low density lipoprotein oxidizability induced by free radicals. *Atherosclerosis* **81**, 175–182.

Balla, G., Jacob, J. S., Eaton, J. W., Belcher, J. D., and Vercellotti, G. M. (1991). Hemin: A possible physiological mediator of low density lipoprotein oxidation and endothelial injury. *Arterioscler. Thromb.* **11**, 1700–1711.

Bedwell, S., Dean, R. T., and Jessup, W. (1989). The action of defined oxygen-centered free radicals on human low-density lipoprotein *Biochem. J.* **262**, 707–712.

Belcher, J. D., Balla, J., Balla, G., Jacobs, D. R. Jr., Gross, M., Jacob, H. S., and Vercellotti, G. M. (1993). Vitamin E, LDL and endothelium: Brief oral vitamin supplementation prevents oxidized LDL-mediated vascular injury in vitro. *Arterioscler. Thromb.* **13**, 1779–1789.

Bendich, A., Machlin, L. J., Scandurra, O., Burton, G. W., and Wayner, D. D. M. (1986). The antioxidant role of vitamin C. *Adv. Free Radical Biol. Med.* **2**, 419–444.

Benedetti, A., Comporti, M., and Esterbauer, H. (1980). Identification of 4-hydroxynonenal as a cytotoxic product originating from the peroxidation of liver microsomal lipids. *Biochim. Biophys. Acta* **620**, 281–296.

Berliner, J. A., Territo, M. C., Sevanian, A., Ramin, S., Kim, J. A., Bamshad, B., Esterson, M., and Fogelman, A. (1990). Minimally modified LDL stimulates monocyte endothelial interactions. *J. Clin. Invest.* **85**, 1260–1266.

Berliner, J. A. and Heinecke, J. (1996). The role of oxidized lipoproteins in atherogenesis. *Free Radical Biol. Med.*, **20**, 707–727.

Boulanger, C. M., Tanner, F. C., Bea, M. L., Hahn, A. W., Werner, A., and Lüscher, T. F. (1992). Oxidized low density lipoproteins induce mRNA expression and release of endothelin from human and porcine endothelium. *Circ. Res.* **70**, 1191–1197.

Bowry, V. W., and Stocker, R. (1993). Tocopherol-mediated peroxidation: The prooxidant effect of vitamin E on the radical-initiated oxidation of human low-density lipoprotein. *J. Am. Chem. Soc.* **115**, 6029–6044.

Bowry, V., Ingold, K. U., and Stocker, R. (1992). Vitamin E in human low-density lipoprotein: When and how this antioxidant becomes a pro-oxidant. *Biochem. J.* **288**, 341–344.

Brown, M. S., and Goldstein, J. L. (1986). A receptor-mediated pathway for cholesterol homeostatis. *Science* **232**, 34–47.

Coffey, M. D., Cole, R. A., Colles, S. M., and Chisolm, G. M. (1995). In vitro cell injury by oxidized low density lipoprotein involves lipid hydroperoxide-induced formation of alkoxyl, lipid, and peroxyl radicals. *J. Clin. Invest.* **96**, 1866–1873.

Cominacini, L., Garbin, U., Cenci, B., Davoli, A., Pasini, C., Ratti, E., Gabiraghi, G., Lo Cascio, V., and Pastorino, A. M. (1991). Predisposition to LDL oxidation during copper-catalyzed oxidative modification and its relation to α-tocopherol content in humans. *Clin. Chim. Acta* **204**, 57–68.

Crawford, D. W., and Blankenhorn, D. H. (1991). Arterial wall oxygenation, oxyradicals, and atherosclerosis. *Atherosclerosis* **89**, 97–108.

Dabbagh, A. J., and Frei, B. (1995). Human suction blister interstitial fluid prevents metal ion-dependent oxidation of low density lipoprotein by macrophages and in cell-free systems. *J. Clin. Invest.* **96**, 1958–1966.

Dieber-Rotheneder, M., Puhl, H., Waeg, H., Striegl, G., and Esterbauer, H. (1991). Effect of oral supplementation with D-α-tocopherol on the vitamin E content of human low density lipoproteins and resistance to oxidation. *J. Lipid Res.* **32**, 1325–1332.

Drake, T. A., Hannani, K., Fei, H. H., Lavi, S., and Berliner, J. A. (1991). Minimally oxidized low-density lipoprotein induces tissue factor expression in cultured human endothelial cells. *Am. J. Pathol.* **138**, 601–607.

Endemann, G., Stanton, L. W., Madden, K. S., Bryant, C. M., White, R. T., and Protter, A. A. (1993). CD36 is a receptor for oxidized low density lipoprotein *J. Biol Chem.* **268**, 11811–11816.

Esterbauer, H., Jürgens, G., Quehenberger, O., and Koller, E. (1987). Autoxidation of human low density lipoprotein: Loss of polyunsaturated fatty acids and vitamin E and generation of aldehydes. *J. Lipid Res.* **28**, 495–509.

Esterbauer, H., Striegl, G., Puhl, H., Oberreither, S., Rotheneder, M., El-Saadani, M., and Jürgens, G. (1989). The role of vitamin E and carotenoids in preventing oxidation of low density lipoprotein. *Ann. N.Y. Acad. Sci.* **570**, 254–267.

Esterbauer, H., Dieber-Rotheneder, M., Waeg, F., Striegl, G., and Jürgens, G. (1990). Biochemical, structural, and functional properties of oxidized low-density lipoprotein. *Chem. Res. Toxicol.* **3**, 77–92.

Esterbauer, H., Dieber-Rotheneder, M., Striegl, G., and Waeg, G. (1991). Role of vitamin E in preventing the oxidation of low-density lipoprotein *Am. J. Clin. Nutr.* **53**(Suppl), 314S–321S.

Esterbauer, H., Gebicki, J., Puhl, H., and Jürgens, G. (1992). The role of lipid peroxidation and antioxidants in oxidative modification of LDL. *Free Radical Biol. Med.* **13**, 341–390.

Flavahan, N. A. (1992). Atherosclerosis of lipoprotein-induced endothelial dysfunction: Potential mechanisms underlying reduction in EDRF/nitric oxide activity. *Circulation* **85**, 1927–1938.

Freeman, M., Ekkel, Y., Rohrer, L., Penman, M., Freedman, N. J., Chisolm, G. M., and Krieger, M. (1991). Expression of type I and type II bovine scavenger receptors in Chinese hamster ovary cells: Lipid droplet accumulation and nonreciprocal cross competition by acetylated oxidized low density lipoprotein. *Proc. Natl. Acad. Sci. U.S.A.* **88**, 4931–4935.

Frei, B. (1995). Cardiovascular disease and nutrient antioxidants: Role of low-density lipoprotein oxidation. *Crit. Rev. Food Sci. Nutr.* **35**, 83–98.

Frei, B., and Gaziano, J. M. (1993). Content of antioxidants, preformed lipid hydroperoxides, and cholesterol as predictors of the susceptibility of human LDL to metal ion-dependent and -independent oxidation. *J. Lipid Res.* **34**, 2135–2145.

Frei, B., and Lynch, S. M. (1994). Effects of combined supplementation with antioxidants on low-density lipoprotein oxidation. *Circulation* **90**, 3119.

Frei, B., Stocker, R., and Ames, B. N. (1988). Antioxidant defenses and lipid peroxidation in human blood plasma. *Proc. Natl. Acad. Sci. U.S.A.* **85**, 9748–9752.

Frei, B., England, L., and Ames, B. N. (1989). Ascorbate is an outstanding antioxidant in human blood plasma. *Proc. Natl. Acad. Sci. U.S.A.* **86**, 6377–6381.

Frei, B., Stocker, R., England L., and Ames, B. N. (1990). Ascorbate: The most effective antioxidant in human blood plasma. *Adv. Exp. Med. Biol.* **264**, 155–163.

Frei, B., Forte, T., Ames, B. N., and Cross, C. E. (1991). Gas-phase oxidants of cigarette smoke induce lipid peroxidation and changes in lipoprotein properties in human blood plasma: Protective effects of ascorbic acid. *Biochem. J.* **277**, 133–138.

Frei, B., Stocker, R., and Ames, B. N. (1992). Small molecule antioxidant defenses in human extracellular fluids. *In* "The Molecular Biology of Free Radical Scavenging Systems" (J. Scandalios, ed.), pp. 23–45. Cold Spring Harbor Laboratory Press, Cold Spring Harbor, NY.

Frostegard, J., Haegerstrand, A., Giglund, M., and Nilsson, J. (1991). Biologically modified LDL increases the adhesive properties of endothelial cells. *Atherosclerosis* **90**, 119–126.

Gaziano, M. J., Manson, J. E., and Hennekens, C. H. (1994). Natural antioxidants and cardiovascular disease: Observational epidemiologic studies and randomized trials. *In* "Natural Antioxidants in Human Health and Disease" (B. Frei, ed.), pp. 387–410. Academic Press, San Diego.

Gieseg, S. P., and Esterbauer, H. (1994). Low density lipoprotein is saturable by pro-oxidant copper. *FEBS Lett.* **343**, 188–194.

Goldstein, J. L., Ho, Y. K., Basu, S. K., and Brown, M. S. (1979). Binding site on macrophages that mediates uptake and degradation of acetylated low density lipoprotein, producing massive cholesterol deposition. *Proc. Natl. Acad. Sci. U.S.A.* **76**, 333–337.

Gotto, A. J., and Farmer, J. A. (1988). Risk factors for coronary artery disease. *In* "Heart Disease: A Textbook of Cardiovascular Medicine," 3rd ed. (E. Braunwald, ed.), pp. 1153–1190. W. B. Saunders, Philadelphia.

Haberland, M. E., Olch, C. L., and Fogelman, A. M. (1984). Role of lysines in mediating interaction of modified low density lipoproteins with the scavenger receptor of human monocyte macrophages. *J. Biol. Chem.* **259,** 11305–11311.

Hajjar, D. P., Farber, I. C., and Smith, S. C. (1988). Oxygen tension within the arterial wall: Relationship to altered bioenergetic metabolism and lipid accumulation. *Arch. Biochem. Biophys.* **262,** 375–380.

Halliwell, B., Wasil, M., and Grootveld, M. (1987). Biologically significant scavenging of the myeloperoxidase-derived oxidant hypochlorous acid by ascorbic acid: Implications for antioxidant protection in the inflamed rheumatoid joint. *FEBS Lett.* **213,** 15–17.

Harats, D., Ben-Naim, M., Dabach, Y., Hollander, G., Havivi, E., Stein, O., and Stein, Y. (1990). Effect of vitamin C and E supplementation on susceptibility of plasma lipoproteins to peroxidation induced by acute smoking. *Atherosclerosis* **85,** 47–54.

Hatta, A., and Frei, B. (1995). Oxidative modification and antioxidant protection of human low-density lipoprotein at high and low oxygen partial pressures. *J. Lipid Res.*, **36,** 2383–2393.

Hazell, L. J., and Stocker, R. (1993). Oxidation of low-density lipoprotein with hypochlorite causes transformation of the lipoprotein into a high-uptake form for macrophages. *Biochem. J.* **290,** 165–172.

Hazell, L. J., van den Berg, J. J., and Stocker, R. (1994). Oxidation of low-density lipoprotein by hypochlorite causes aggregation that is mediated by modification of lysine residues rather than lipid oxidation. *Biochem. J.* **302,** 297–304.

Heinecke, J. W., Baker, L., Rosen, H., and Chait, A. (1986). Superoxide mediated modification of low density lipoprotein by human arterial smooth muscle cells in culture. *J. Clin. Invest.* **77,** 757–761.

Heinecke, J. W., Rosen, H., Suzuki, L. A., and Chait, A. (1987). The role of sulfur-containing amino acids in superoxide production and modification of low density lipoproteins by arterial smooth muscle cells. *J. Biol. Chem.* **62,** 10098–10103.

Henriksen, T., Mahoney, E. M., and Steinberg, D. (1981). Enhanced macrophage degradation of low density lipoprotein previously incubated with cultured endothelial cells: Recognition by receptor for acetylated low density lipoproteins. *Proc. Natl. Acad. Sci. U.S.A.* **78,** 6499–6503.

Hunter, E. P. L., Desrosiers, M. R., and Simic, M. G. (1989). The effect of oxygen, antioxidants, and superoxide radical on tyrosine phenoxyl radical dimerization. *Free Radical Biol. Med.* **6,** 581–585.

Ingold, K. U., Bowry, V. W., Stocker, R., and Walling, C. (1993). Autoxidation of lipids and antioxidation by α-tocopherol and uniquinol in homogeneous solution and in aqueous dispersions of lipids: Unrecognized consequences of lipid particle size as exemplified by oxidation of human low density lipoprotein. *Proc. Natl. Acad. Sci. U.S.A.* **90,** 45–49.

Iwatsuki, M., Kiki, E., Stone, D., and Darley-Usmar, V. M. (1995). α-Tocopherol mediated peroxidation in the copper (II) and met myoglobin induced oxidation of human low density lipoprotein: The influence of lipid hydroperoxides. *FEBS Lett.* **360,** 271–276.

Jessup, W., Rankin, S. M., De Whalley, C. V., Hoult, J. R. S., Scott, J., and Leake, D. S. (1990). Alpha-tocopherol consumption during low-density-lipoprotein oxidation. *Biochem. J.* **265,** 399–405.

Jialal, I., and Grundy, S. M. (1991). Preservation of the endogenous antioxidants in low density lipoprotein by ascorbate but not probucol during oxidative modification. *J. Clin. Invest.* **87,** 597–601.

Jialal, I., and Grundy, S. M. (1992). Effect of dietary supplementation with alpha-

tocopherol on the oxidative modification of low density lipoprotein. *J. Lipid Res.* **33**, 899–906.
Jialal, I., and Grundy, S. M. (1993). Effect of combined supplementation with α-tocopherol, ascorbate, and beta carotene on low-density lipoprotein oxidation. *Circulation* **88**, 2780–2786.
Jialal, I., Vega, G. L., and Grundy, S. M. (1990). Physiologic levels of ascorbate inhibit oxidative modification of low density lipoprotein. *Atherosclerosis* **82**, 185–191.
Jialal, I., Fuller, C. J., and Huet, B. A. (1995). The effect of α-tocopherol supplementation on LDL oxidation: A dose-response study. *Arterioscler. Thromb. Vasc. Biol.* **15**, 190–198.
Kagan, V. E., Serbinova, E. A., Forte, T., Scita, G., and Packer, L. (1992). Recycling of vitamin E in human low density lipoproteins. *J. Lipid Res.* **33**, 385–397.
Keaney, J. F. Jr., and Frei, B. (1994). Antioxidant protection of low-density lipoprotein and its role in the prevention of atherosclerotic vascular disease. *In* "Natural Antioxidants in Human Health and Disease" (B. Frei, ed.), pp. 303–351. Academic Press, San Diego.
Keaney, J. F. Jr., Gaziano, J. M., Xu, A., Frei, B., Curran-Celentano, J., Shwaery, G. T., Loscalzo, J., and Vita, J. A. (1993). Dietary antioxidants preserve endothelium-dependent vessel relaxation in cholesterol-fed rabbits. *Proc. Natl. Acad. Sci. U.S.A.* **90**, 11880–11884.
Khoo, J. C., Miller, E., Pio, F., Steinberg, D., and Witztum, J. L. (1992). Monoclonal antibodies against LDL further enhance macrophage uptake of LDL aggregates. *Arterioscler. Thromb.* **12**, 1258–1266.
Kim, J. A., Territo, M. C., Wayner, E., Carlos, T. M., Parhami, F., Smith, C. W., Haberland, M. E., Fogelman, A. M., and Berliner, J. A. (1994). Partial characterization of leukocyte binding molecules on endothelial cells induced minimally oxidized LDL. *Arterioscler. Thromb.* **14**, 427–433.
Krieger, M., and Herz, J. (1994). Structures and functions of multiligand lipoprotein receptors: Macrophage scavenger receptors and LDL receptor-related protein (LRP). *Annu. Rev. Biochem.* **63**, 601–637.
Kritharides, L., Jessup, W., and Dean, R. T. (1995). Macrophages require both iron and copper to oxidize low-density lipoprotein in Hanks' balanced salt solution. *Arch. Biochem. Biophys.* **323**, 127–136.
Kuzuya, M., Yamada, K., Hayashi, T., Funaki, C., Naito, M., Asai, K., and Kuzuya, F. (1992). Role of lipoprotein-copper complex in copper catalyzed-peroxidation of low-density lipoprotein. *Biochim. Biophys. Acta* **1123**, 334–341.
Latron, Y., Chautan, M., Anfosso, F., Alessi, M. C., Nalbone, G., Lafont, H., and Juhan-Vague, I. (1991). Stimulating effect of oxidized low density lipoproteins on plasminogen activator inhibitor-1 synthesis by endothelial cells. *Arterioscler. Thromb.* **11**, 1821–1829.
Leake, D. S., and Rankin, S. M. (1990). The oxidative modification of low-density lipoproteins by macrophages. *Biochem. J.* **270**, 741–748.
Lynch, S. M., and Frei, B. (1993). Mechanisms of copper- and iron-dependent oxidative modification of human low-density lipoprotein. *J. Lipid Res.* **34**, 1745–1753.
Lynch, S. M., and Frei, B. (1995). Reduction of copper, but not iron, by human low density lipoprotein (LDL). Implications for metal ion-dependent oxidative modification of LDL. *J. Biol. Chem.* **270**, 5158–5163.
Lynch, S. M., Morrow, J. D., Roberts, L. J. II, and Frei, B. (1994). Formation of non-cyclooxygenase-derived prostanoids (F_2-isoprostanes) in plasma and low density lipoprotein exposed to oxidative stress in vitro. *J. Clin. Invest.* **93**, 998–1004.

Lynch, S. M., Gaziano, J. M., and Frei, B. (1995). Ascorbic acid and atherosclerotic cardiovascular disease. In "Ascorbic Acid: Biochemistry and Biomedical Cell Biology" (J. R. Harris, ed.), pp. 331–367. Plenum Press, New York.

Ma, Y. S., Stone, W. L., and LeClair, I. O. (1994). The effects of vitamin C and urate on the oxidation kinetics of human low-density lipoprotein. *Proc. Soc. Exp. Biol. Med.* **206**, 53–59.

Miller, N. J., Paganga, G., Wiseman, S., Van Nielen, W., Tijburg, L., Chowienczyk, P., and Rice-Evans, C. A. (1995). Total antioxidant activity of low density lipoproteins and the relationship with α-tocopherol status. *FEBS Lett.* **365**, 164–166.

Minotti, G., and Aust, S. D. (1987). The role of iron in the initiation of lipid peroxidation. *Chem. Phys. Lipids* **44**, 191–208.

Moore, K. P., Darley-Usmar, V., Morrow, J., and Roberts, L. J. II. (1995). Formation of F_2-isoprostanes during oxidation of human low-densityh lipoprotein and plasma by peroxynitrite. *Circ. Res.* **77**, 335–341.

Morel, D. W., DiCorleto, P. E., and Chisolm, G. M. (1984). Endothelial and smooth muscle cells alter low density lipoprotein in vitro by free radical oxidation. *Arteriosclerosis* **4**, 357–364.

Naseem, K. M., Goodall, A. H., and Bruckdorfer, K. R. (1993). The effects of native and oxidised low density lipoproteins on platelet activation. *Biochem. Soc. Trans.* **21**, 140S.

National Research Council. (1989). Recommended Dietary Allowances, 10th ed. National Academy Press, Washington, DC.

Neuzil, J., and Stocker, R. (1994). Free and albumin-bound bilirubin are efficient co-antioxidants for α-tocopherol, inhibiting plasma and low density lipoprotein lipid peroxidation. *J. Biol. Chem.* **269**, 16712–16719.

Nishikimi, M. (1975). Oxidation of ascorbic acid with superoxide anion generated by the xanthine-xanthine oxidase system. *Biochem. Biophys. Res. Commun.* **63**, 463–468.

Ottnad, E., Parthasarathy, S., Sambrano, G. R., Ramprasad, M. P., Quehenberger, O., Kondratenko, N., Green, S., and Steinberg, D. (1995). A macrophage receptor for oxidized low density lipoprotein distinct from the receptor for acetyl low density lipoprotein: Partial purification and role in recognition of oxidatively damaged cells. *Proc. Natl. Acad. Sci. U.S.A.* **92**, 1391–1395.

Palinski, W., Rosenfeld, M. E., Ylä-Herttuala, S. Gurtner, G. C., Socher, S. S., Butler, S. W., Parthasarathy, S., Carew, T. E., Steinberg, D., and Witztum, J. L. (1989). Low density lipoprotein undergoes oxidative modification in vivo. *Proc. Nat. Acad. Sci. U.S.A.* **86**, 1372–1376.

Parhami, F., Fang, Z. T., Fogelman, A. M., Andalibi, A., Territo, M. C., and Berliner, J. A. (1993). Minimally modified low density lipoprotein-induced inflammatory responses in endothelial cells are mediated by cyclic adenosine monophosphate. *J. Clin. Invest.* **92**, 471–478.

Parasarathy, S., Printz, D. J., Boyd, D., Joy, L., and Steinberg, D. (1986). Macrophage oxidation of low density lipoprotein generates a modified form recognized by the scavenger receptor. *Arteriosclerosis* **6**, 505–510.

Princen, H. M., van Poppel, G., Vogelezang, C., Buytenhek, R., and Kok, F. J. (1992). Supplementation with vitamin E but not β-carotene in vivo protects low density lipoprotein from lipid peroxidation in vitro: Effect of cigarette smoking. *Arterioscler. Thromb.* **12**, 554–562.

Princen, H. M. G., van Duyvenvoorde, W., Buytenhek, R., van der Laarse, A., van Poppel, G., Leuven, J. A. G., and van Hinsbergh, V. W. M. (1995). Supplementation with low

doses of vitamin E protects LDL from lipid peroxidation in men and women. *Arterioscler. Thromb. Vasc. Biol.* **15**, 325–333.
Quinn, M. T., Parthasarathy, S., Fong, L. G., and Steinberg, D. (1987). Oxidatively modified low density lipoproteins: A potential role in recruitment and retention of monocyte/macrophages during atherogenesis. *Proc. Natl. Acad. Sci. U.S.A.* **84**, 2995–2998.
Quinn, M. T., Parthasarathy, S., and Steinberg, D. (1988). Lysophosphatidylcholine: A chemotactic factor for human monocytes and its potential rote in atherogenesis. *Proc. Natl. Acad. Sci. U.S.A.* **84**, 1372–1376.
Reaven, P. D., and Witztum, J. L. (1993). Comparison of supplementation of RRR-α-tocopherol and racemic α-tocopherol in humans: Effects of lipid levels and lipoprotein susceptibility to oxidation. *Arterioscler, Thromb.* **13**, 601–608.
Reaven, P., Parthasarathy, S., Grasse, B. J., Miller, E., Almazan, F., Mattson, F. H., Khoo, J. C., Steinberg, D., and Witztum, J. L. (1991). Feasibility of using an oleate-rich diet to reduce the susceptibility of low-density lipoprotein to oxidative modification in humans. *Am. J. Clin. Nutr.* **54**, 701–706.
Reaven, P. D., Khouw, A., Beltz, W. F., Parthasarathy, S., and Witztum, J. L. (1993). Effect of dietary antioxidant combinations in humans: Protection of LDL by vitamin E but not beta-carotene. *Arterioscler. Thromb.* **13**, 590–600.
Retsky, K. L., and Frei, B. (1995). Vitamin C prevents metal ion-dependent initiation and propagation of lipid peroxidation in human low density lipoprotein. *Biochim. Biophys. Acta.* **1257**, 279–287.
Retsky, K. L., Freeman, M. W., and Frei, B. (1993). Ascorbic acid oxidation product(s) protect human low density lipoprotein against atherogenic modification: Anti- rather er than prooxidant activity of vitamin C in the presence of transition metal ions. *J. Biol. Chem.* **268**, 1304–1309.
Rifici, V. A., and Khachadurian, A. K. (1993). Dietary supplementation with vitamins C and E inhibits in vitro oxidation of lipoproteins. *J. Am. Coll. Nutr.* **12**, 631–637.
Samuni, A., Aronovitch, J., Godinger, D., Chevion, M., and Czapski, G. (1983). On the cytotoxicity of vitamin C and metal ions: A site-specific Fenton mechanism. *Eur. J. Biochem.* **137**, 119–124.
Sato, K., Niki, E., and Shimasaki, H. (1990). Free radical-mediated chain oxidation of low density lipoprotein and its synergistic inhibition by vitamin E and C. *Arch. Biochem. Biophys.* **279**, 402–405.
Savenkova, M. I., Mueller, D. M., and Heinecke, J. W. (1994). Tyrosyl radical generated by myeloperoxidase is a physiological catalyst for the initiation of lipid peroxidation in low density lipoprotein. *J. Biol. Chem.* **269**, 20394–20400.
Schwartz, C. J., and Valente, A. J. (1994). The pathogenesis of atherosclerosis. In "Natural Antioxidants in Human Health and Disease" (B. Frei, ed.), pp. 287–302. Academic Press, San Diego.
Shaish, A., Daugherty, A., O'Sullivan, F., Schonfeld, G., and Heinecke, J. W. (1995). Beta-carotene inhibits atherosclerosis in hypercholesterolemic rabbits. *J. Clin. Invest.* **96**, 2075–2082.
Smith, D., O'Leary, V. J., and Darley-Usmar, V. M. (1993). The role of α-tocopherol as a peroxyl radical scavenger in human low density lipoprotein. *Biochem. Parmacol.* **45**, 2195–2201.
Sparrow, C. P., and Olszewski, J. (1993). Cellular oxidation of low density lipoprotein is caused by thiol production in media containing transition metal ions. *J. Lipid Res.* **34**, 1219–1228.

Sparrow, C. P., Parthasarathy, S., Steinberg, D. (1989). A macrophage receptor that recognizes oxidized low density lipoprotein but not acetylated low density lipoprotein. *J. Biol. Chem.* **264,** 2599–2604.

Stait, S. E., and Leake, D. S. (1994). Ascorbic acid can either increase or decrease low density lipoprotein modification. *FEBS Lett.* **341,** 263–267.

Stanton, L. W., White, R. T., Bryant, C. M., Protter, A. A., and Endemann, G. (1992). A macrophage F_c receptor for IgG is also a receptor for oxidized low density lipoprotein. *J. Biol. Chem.* **267,** 22446–22451.

Steinberg, D., Parthasarathy, S., Carew, T. E., Khoo, J. C., and Witztum, J. L. (1989). Beyond cholesterol: Modifications of low-density lipoprotein that increase its atherogenicity. *New Engl. J. Med.* **320,** 915–924.

Steinbrecher, U. P. (1987). Oxidation of human low-density lipoprotein results in derivatization of lysine residues of apolipoprotein B by lipid peroxide decomposition products. *J. Biol. Chem.* **262,** 3603–3608.

Steinbrecher, U. P. (1988). Role of superoxide in endothelial-cell modification of low-density lipoproteins. *Biochim. Biophys. Acta* **959,** 20–30.

Steinbrecher, U. P., Parthasarathy, S., Leake, D. S., Witztum, J. L., and Steinberg, D. (1984). Modification of low density lipoprotein by endothelial cells involves lipid peroxidation and degradation of low density lipoprotein phospholipids. *Proc. Natl. Acad. Sci. U.S.A.* **81,** 3883–3887.

Stocker, R., and Frei, B. (1991). Endogenous antioxidant defences in human blood plasma. *In* "Oxidative Stress: Oxidants and Antioxidants" (H. Sies, ed.), pp. 213–243. AcaademicPress,London.

Stocker, R., Bowry, V. W., and Frei, B. (1991). Ubiquinol-10 protects human low density lipoprotein more efficiently against lipid peroxidation than does α-tocopherol. *Proc. Natl. Acad. Sci. U.S.A.* **88,** 1646–1650.

Sugiyama, S., Kugiyama, K., Ohgushi, M., Fujimoto, K., and Yasue, H. (1994). Lysophosphatidylcholine in oxidized low-density lipoprotein increases endothelial susceptibility to polymorphonuclear leukocyte-induced endothelial dysfunction in porcine coronary arteries: Role of protein kinase C. *Circ. Res.* **74,** 565–575.

Suits, A. G., Chait, A., Aviram, M., and Heinecke, J. W. (1989). Phagocytosis of aggregated lipoprotein by macrophages: Low density lipoprotein receptor-dependent foam-cell formation. *Proc. Nat. Acad. Sci. U.S.A.* **86,** 2713–2717.

Thorin, E., Hamilton, C. A., Dominiczak, M. H., and Reid, J. L. (1994). Chronic exposure of cultured bovine endothelial cells to oxidized LDL abolishes prostacyclin release. *Arterioscler. Thromb.* **14,** 453–459.

Uchida, K., and Kawakishi, S. (1990). Site-specific oxidation of angiotensin I by copper(II) and L-ascorbate: Conversion of histidine residues to 2-imidazolones. *Arch. Biochem. Biophys.* **283,** 20–26.

van Hinsbergh, V. W. M., Scheffer, M., Havekes, L., and Kempen, H. J. M. (1986). Role of endothelial cells and their products in the modification of low-density lipoproteins. *Biochim. Biophys. Acta* **878,** 49–64.

Weis, J. R., Pitas, R. E., Wilson, B. D., and Rodgers, G. M. (1991). Oxidized low-density lipoprotein increases cultured human endothelial cell tissue factor activity and reduces protein C activation. *FASEB J.* **5,** 2459–2465.

Welch, G., and Loscalzo, J. (1994). Nitric oxide and the cardiovascular system. *J. Cardiovasc. Surg.* **9,** 361–371.

Yoshida, Y., Tsuchiya, J., and Niki, E. (1994). Interaction of α-tocopherol with copper and its effect on lipid peroxidation. *Biochim. Biophys. Acta* **1200,** 85–92.

Antioxidant Vitamins and Human Immune Responses

ADRIANNE BENDICH

Human Nutrition Research, Hoffmann-La Roche Inc., Paramus, New Jersey 07652

I. Introduction
II. Free Radicals
III. Essential Micronutrients with Antioxidant Activities
IV. Immune Functions
   A. Assessment of Immune Functions—Oxidative Stress and Antioxidant Status
   B. Delayed Hypersensitivity
   C. Vaccine Titers
   D. Tumor Cell Killing
V. Free Radicals and Immune Cell Function
   A. Effects of Different Fatty Acids
   B. Effects of Immune-Generated Oxidants
   C. Effects of Antioxidant Deficiency
VI. Clinical Examples of the Interaction of Free Radicals, Antioxidants, and Immune Function
   A. Rheumatoid Arthritis
   B. Aging
   C. Cigarette Smoking
VII. Conclusions
References

I. Introduction

In the last decade, there has been growing evidence that oxidative stress impairs immune responses and that antioxidant supplementation can reverse many aspects of oxidant-mediated immune suppression.

The objectives of this review are to (1) examine the role of free radicals and other reactive oxygen species in the immune system, (2) discuss the relevant immune parameters for assessment of the effects of antioxidants, and finally (3) by use of three examples, provide a synopsis of recent clinical findings. The three examples include an autoimmune disease (rheumatoid arthritis), the effects of aging, and the effects of an environmental agent (cigarette smoke). These examples are chosen because of the relatively large number of clinical studies in these areas that address the interactions of immune function and oxidant–antioxidant balance.

It is hoped that this review increases the awareness of the clinical data, which consistently show an improvement in immune responses when antioxidant status is improved. The health care consequences of these data should not be minimized.

II. Free Radicals

Free radicals are highly reactive molecules with one or more unpaired electrons. Free radicals are generated during cellular metabolism, can be ingested or inhaled as environmental pollutants, or can be generated during the metabolism of certain drugs or xenobiotics (Halliwell, 1995; Machlin and Bendich, 1987). Oxygen-containing free radicals such as the superoxide and hydroxyl radicals are associated with damage to cell structures and, consequently, human diseases, including those affecting immune functions. In addition to these free radicals, there are other reactive oxygen-containing species that can compromise health. Two of the most common, singlet oxygen and hydrogen peroxide, have also been linked to immune depression.

III. Essential Micronutrients with Antioxidant Activities

Antioxidants interfere with the production of free radicals and/or inactivate them once they are formed. There are three enzymes that have antioxidant capacity and contain essential minerals. Two are types of superoxide dismutases: a manganese-containing enzyme and a copper–zinc-containing enzyme. The third enzyme, catalase, an iron-containing enzyme, catalyzes the decomposition of hydrogen peroxide. Selenium is an essential component of glutathione peroxidase, important in the decomposition of hydrogen peroxide and lipid peroxides, termination products of free radical attack on lipids (Machlin and Bendich, 1987). Although not discussed in this review, it is important to mention the exciting data that suggests that the virulence of certain viral pathogens is inversely related to the selenium (and/or vitamin E) status of the host (Beck *et al.*, 1995). In animal models, selenium deficiency resulted in the transformation of a nonvirulent strain of coxsackievirus to the virulent state (Beck *et al.*, 1994a,b). This clinically relevant new research area warrants close attention.

Three essential micronutrients can directly interfere with the propagation stage of free radical generation as well as scavenge free radicals. Vitamin E (α-tocopherol), the major lipid-soluble antioxidant present in all cellular membranes, protects against lipid peroxidation and prevents the loss of membrane fluidity. Vitamin E has been characterized as the most critical antioxidant in blood (Burton *et al.*, 1983). Vitamin C (ascorbic acid), is water soluble and, along with vitamin E, can quench free radicals as well as singlet oxygen (Frei, 1991). Ascorbate can also regenerate the reduced, antioxidant form of vitamin E (Bendich *et al.*, 1986).

Recent work has shown that β-carotene, a pigment found in all photosynthetic plants, is an efficient quencher of singlet oxygen and can function as an antioxidant (Krinsky, 1989). Carotenoid-rich diets are consistently associated with reduced risk of many cancers (Ziegler and Subar, 1991). β-Carotene is the major dietary carotenoid precursor of vitamin A. Vitamin A cannot quench singlet oxygen and has less antioxidant activity than the other nutrients discussed; however, its importance in clinical research (Bendich, 1994) and for the immune system (Bendich, 1993) is becoming well recognized.

IV. Immune Functions

The cells, tissues, organs, and products of the immune system are found throughout all of the other tissues and organs of the body. The primary function of the immune system is to maintain the integrity of the body, to protect against invasion by pathogens, to remove damaged or aged cells, and to seek out and destroy altered cells with the potential to cause cancer (Chandra and Kumari, 1994). The immune system also has the capacity to overreact or, in contrast, lose its ability to respond appropriately.

A. Assessment of Human Immune Responses—Oxidative Stress and Antioxidant Status

In this review, focus is placed on human studies that have used relevant assays to measure the effects of free radicals and/or antioxidants on immune responses. For the readers' information, Chew (1995) has reviewed the importance of antioxidant vitamins on production animals' immune function.

1. *Noninvasive Measures*

Noninvasive measures of immune function are particularly important in large epidemiological trials for assessment of easily measured parameters and are often used to determine health care cost–benefit ratios. These assessments include infection rates, number of days of fever, days of use of antibiotics, days of hospitalization, tonsil size in young children, and number of sneezes, coughs, and nose blows per day. Noninvasive measures can also be used to determine oxidative stress; the most commonly used is the measurement of the oxidation products pentane and ethane in the breath. Dietary recall is the most commonly used noninvasive measure of intake of antioxidants.

2. *Invasive Measures*

Invasive measures are defined as those that require either the non-oral administration of agents to the subject or the removal of blood or other internal body fluids from the subject. Invasive measures of immune function include delayed-type hypersensitivity (DTH) skin test responses; antibody titers to vaccines; serum concentrations of acute-phase proteins and virus concentrations; and determination of white blood cell functions such as natural killer (NK) cell function, macrophage and neutrophil phagocytic functions, and lymphocyte proliferation as well as assays of interleukins and cytokines (Bendich, 1995).

In addition, the fluorescent-activated cell sorter has provided an accurate and rapid means for determining subsets of immune cells, their receptors, and their state of activation. Serum concentrations of lipid peroxides and other oxidative products of proteins, DNA, and carbohydrates are used to assess levels of oxidative stress. Serum concentrations of the antioxidant micronutrients are determined to evaluate antioxidant status. Three assays require a more detailed description because these are the measures used most frequently in the clinical studies discussed later.

B. Delayed Hypersensitivity

DTH skin test responses are mainly a reflection of the memory T-cell function of an individual. The DTH response is invoked by reexposure to an antigen; usually seven antigens are tested at the same time. The magnitude of the immune response is reflected by the summation of the number of responses and the diameter of the redness surrounding each site of antigen placement (induration), usually under the skin of the forearm, assessed 24–48 h after exposure. In addition to the memory T-cell response, DTH also is dependent upon the interaction of antigen-presenting cells, effector T cells, cytokine production, and inflammatory responses at the site of antigen exposure. A vigorous DTH response reflects the effective interactions of all cells and factors necessary to mount a similar immune response should an adult individual be exposed to most pathogens in the environment. The DTH response is also similar to the response seen postvaccination to a common pathogen. Thus DTH is used as an indirect measure of an individual's capacity to mount an immune response to infection or immune-related disease. Clinically, DTH is equally important as an index of immune compromise; lack of DTH response to all seven antigens is defined as anergy and is reflective of increased risk of morbidity and mortality, especially in the elderly (Wayne et al., 1990; Christou, 1990).

C. Vaccine Titers

Responses to vaccines are determined by the vigor of the antibody response to the viral or bacterial antigen challenge. It is well documented that the elderly have reduced antibody titers to flu vaccines as they age, which may be related to the overall depression of T-cell functions seen in aging (Chandra, 1992).

It is of interest to note that the infectivity of viruses such as influenza (flu) may be dependent upon the antioxidant status of the individual.

D. Tumor Cell Killing

The three types of immune cells that have the capacity to lyse tumor cells are cytotoxic T lymphocytes, macrophages, and NK cells. In addition, factors that are secreted by immune cells, including free radicals and oxidation products, can lyse tumor cells. Assessment of tumor cell lysis usually involves the separation of peripheral white blood cells (PBCs) from other blood components and exposure of the individual's PBCs to radiolabeled tumor cells in culture. The tumor cells' destruction is reflected by the radioactivity released from the labeled tumor cells following their lysis by NK or other tumor-killing cells (Bendich, 1990).

V. Free Radicals and Immune Cell Function

There are numerous links between free radical reactions and immune cell functions. White blood cell membranes, as all cellular membranes, are composed of lipids containing saturated and unsaturated fatty acids. Unsaturated bonds in fatty acids are highly susceptible to free radical attack, one consequence of which is to adversely affect the integrity of the cell's membranes. For instance, oxygen-containing radicals and the products of their reactions have been shown to decrease the fluidity of white blood cell membranes (reviewed in Baker and Meydani, 1994) and synovial fluids (Merry et al., 1989), consequently reducing their function.

A. Effects of Different Fatty Acids

Fluidity is, in part, dependent upon the degree of unsaturation and the classes of fatty acids in cellular membranes. As the dietary level of polyunsaturated fatty acids is increased, the potential for free radical-mediated membrane lipid peroxidation is also increased. Loss of membrane fluidity has been directly related to the decreased ability of lymphocytes to respond to challenges to the immune system (Bendich, 1990). The types of fatty acids incorporated into leukocyte membranes also affect their activities. The synthesis of immunosuppressive prostaglandins is increased as the level of dietary vegetable oils, major sources of ω-6 fatty acids, is increased (Alexander, 1995). Fish oils, which contain a relatively high concentration of ω-3 fatty acids, increase the requirement for antioxidants. However, fish oils, which tend to be anti-inflammatory, do not affect immune responses in the same manner as vegetable oils (Meydani et al., 1988).

In clinical studies, the use of fish oil supplements has been found to decrease interleukin-1 and -2 production and lymphocyte proliferation (reviewed in Kramer et al., 1991). In a placebo-controlled, double-blind study, healthy adult males were given 15 g/day of either placebo oil or fish oil for 10 weeks. Lymphocyte blastogenic responses were significantly depressed in the fish oil group. During the next 8 weeks, the fish oil group was also given 200 IU/day of vitamin E and the control group was given a matched placebo. Lymphocyte blastogenic responses of those given vitamin E were restored to the level seen in the fish oil placebo group. Kramer et al., (1991) strongly recommended that subjects using fish oil supplements consider taking a vitamin E supplement.

B. Effects of Immune-Generated Oxidants

In addition to the general effects of fatty acid composition on immune cell function, the most abundant circulating white blood cell, the neutrophil, utilizes reactive oxygen species to kill invading organisms. When stimulated, neutrophils have the capacity to take up molecular oxygen and generate oxygen-containing free radicals and other reactive molecules. This is often called the oxidative burst. Free radicals and singlet oxygen, along with other reactive molecules, can kill bacterial pathogens. Neutrophils can also generate highly toxic halogenated molecules (e.g., hypochlorous acid) when the myeloperoxidase halide enzyme system is activated during the oxidative burst. The halogenated species can also lyse the phagocytized pathogen (Anderson et al., 1990). Another reactive oxygen species, peroxynitrite, has been identified as an important component in immune cell killing of pathogens (Nath et al., 1988; Beckman and Crow, 1993).

Under normal circumstances, the reactive oxygen species generated by neutrophils and other immune cells are used for control of infection. However, in circumstances of chronic activation, the reactive oxygen products can result in destruction of normal tissue as seen in rheumatoid arthritis (discussed later) (Panayi et al., 1992; Kaur and Halliwell, 1994).

C. Effects of Antioxidant Deficiency

One of the most obvious factors that could reduce antioxidant status is an induced dietary deficiency. Jacob et al. (1991) initiated a clinical trial that utilized a metabolic ward where all meals were prepared to contain >80% of the recommended daily allowance (RDA) for all micronutrients (other than vitamin C) and macronutrients. They examined the effects of marginal vitamin C deficiency on immune and other parameters in healthy males during the study, which lasted 92 days. Serum, white blood cell, and sperm vitamin C levels were significantly reduced when the diet contained only 5, 10, or 20 mg of vitamin C for 2 months. DTH responses to 7 antigens were also significantly depressed during the period of low vitamin C intake. In fact, when subjects initially consumed 250 mg/day of vitamin C, they responded to 3.3 of 7 antigens; this response was reduced to less than 1 antigen after only 1 month at 5 mg of vitamin C per day. Even when intakes were increased back to 250 mg/day for 1 month, the average number of DTH responses did not increase above 1 of 7 antigens. The robustness of the response (the diameter of the induration) was 35 mm at baseline; dropped to 11 mm when vitamin C intakes were 5, 10, or 20 mg for 2 months; and increased to half of the initial level (17 mm) when either 60 mg (the current RDA) or 250 mg of vitamin C was consumed daily for 1 month.

In addition to DTH responses, an index of oxidative damage to sperm DNA was also measured; these levels were significantly increased when vitamin C intakes were low, and the baseline levels were not reached even after 1 month of intake of 250 mg of vitamin C. However, the serum, white blood cell, and semen vitamin C concentrations did return to the baseline levels when vitamin C intake was increased to 250 mg/day.

This carefully controlled study clearly demonstrates the importance of the balance between antioxidant status, oxidant levels, and immune response. This study is especially noteworthy since all other dietary antioxidants, such as vitamin E and β-carotene, were provided throughout the study at RDA levels and only one antioxidant, vitamin C, was reduced.

VI. Clinical Examples of the Interactions of Free Radicals, Antioxidants, and Immune Function

Three examples of immune dysfunction associated with increased oxidative stress are discussed in detail because they clearly show the interactions between the immune system and the oxidant–antioxidant status of the individual. The first example, rheumatoid arthritis, is an autoimmune disease of unknown cause. The second area discussed is the decline of immune function associated with the aging process. The last example involves the effects of an environmental factor, cigarette smoke, on immune function.

A. Rheumatoid Arthritis

In the United States, approximately 2.5 million adults have rheumatoid arthritis (RA), which is characterized by a chronic inappropriate immune response in articular joints, resulting in inflammation and destruction of joint tissues. RA is thought to begin when a substance (an antigen) triggers an inappropriate immune response in the articular joints (e.g., elbow, knee, shoulder). The responsible factor is unknown, but viruses, nonbiodegradable products of bacteria, or antibodies directed against structures within the joint have all been implicated. At some point, the initial insult and inflammation becomes a chronic process; the change from acute to chronic inflammation appears to result from a specific immune reaction directed at the persisting antigen (Zvaifler, 1988). The rate of progression of the disease depends on both the intensity and the duration of the inappropriate immune responses (Harris, 1985; Aho et al., 1991).

At the cellular level, the earliest recorded change during the days to weeks following the first symptoms of RA is damage to the lining of the small blood vessels in the joints. At the same time, infiltration of the fluid-filled joint space (synovium) with inflammatory neutrophils occurs and structural changes begin to appear in the synovium, forming an inflamed layer called a pannus. The synovial fluid of the inflamed rheumatoid joint also contains numerous neutrophils (Tarp, 1994). Oxidative products reduce the viscosity of the synovial fluid by oxidizing lipids in synovial fluid, making them more hydrogenated and thereby hindering joint movement further (Halliwell, 1995). The longer term consequences of these joint changes include erosion of cartilage, bone, ligaments, and tendons in the area (Zvaifler, 1988).

Several other components of the immune system contribute to the development of RA. B lymphocytes produce antibodies that form the components of rheumatoid factors. Rheumatoid factors are immunoglobulins (IgM) that inappropriately react with other immunoglobulins (IgG) to form immune complexes. T lymphocytes also generate inappropriate immune responses that result in the increased production of inflammatory cytokines and interferons. Neutrophils and macrophages synthesize and release free radicals, prostaglandins, and immunostimulatory molecules, all of which enhance the inflammatory response in RA. Neutrophils appear to be the major mediators of tissue-destructive events in RA (Harris, 1985; Azzini *et al.*, 1995; Heliovaara *et al.*, 1994).

Some of the adverse immunological consequences result from the alteration of antibodies by free radical reactions. Lunec and Hill (1984) reported that not only did certain immunoglobulins undergo free radical denaturation, but the altered immunoglobulins stimulated neutrophils to generate free radicals.

Animal models have documented the increased production of pentane, an end product of lipid peroxidation, in arthritic animals; vitamin E supplementation lowered the pentane exhalation as well as joint swelling (Tappel and Summerfield, 1984). Humad *et al.* (1988) showed that breath pentane concentrations correlated with the severity of symptoms in RA patients. Coltro *et al.* (1987) found that clinical improvement was associated with a significant decrease in breath pentane in a separate group of RA patients.

Kowsari *et al.* (1983) assessed the baseline nutritional status of arthritis patients in a prospective study. Data indicated that many patients were at least marginally inadequate in selected nutrients, including vitamins E and C. Thurnham *et al.* (1987) conducted a study to assess tocopherol and ascorbic acid antioxidant capacity in RA patients by utilizing the total radical-trapping antioxidant parameter ("Trap") assay. A high Trap value indicates a good antioxidant status. Twenty patients with RA were matched by age and sex to 200 healthy controls. The Trap values in the patients with RA were significantly lower than those in the control subjects. Serum concentrations of tocopherol and ascorbic acid were also significantly lower in the patients compared to the controls.

Heliovaara *et al.* (1994), in a prospective case–control study, found that individuals who developed RA had a significantly lower antioxidant index, which resulted in an eight-fold higher risk of RA. The antioxidant index consisted of a comparison of serum levels of vitamin E, β-carotene, and selenium between the lowest and highest tertiles in cases and controls. The authors suggest that the low antioxidant status could reflect either low dietary intake or increased demand as a result of increased oxidative stress in RA.

With regard to intervention studies, vitamin E has been examined to the greatest extent in patients with RA and other arthritic conditions. Between 1985 and 1994, there were seven studies of the effects of vitamin E supplementation in patients with RA, osteoarthritis, and other related arthritic conditions (Bendich and Cohen, 1996). The data from the three studies in RA patients suggest that vitamin E at high doses (~1500 IU/day) but not doses near dietary intake levels (~6 IU/day) reduced pain (Kolarz *et al.*, 1990; Tulleken *et al.*, 1990; Langley *et al.*, 1983). Vitamin E supplementation, at levels from 100 to about 600 IU/day, produced significant pain relief in the other four studies. There was also a consistent report of vitamin E-related anti-inflammatory activity (Bendich and Cohen, 1996).

B. Aging

The cumulative effects of free radical damage throughout the life span are graphically seen in the pigmented age spots of the elderly, which are a consequence of lipid oxidation. Oxidative damage to the lens of the eye is associated with cataract formation, and increased concentrations of free radical-mediated peroxides are seen in several other tissues in the aged. Cumulative oxidative damage visible in the skin and cataracts of the elderly can also be seen systemically in deposits of oxidized lipids in blood vessel and organs. The overall increased oxidative stress associated with aging is thought to also adversely affect many aspects of immune responses (Bendich, 1995).

Immune function impacts the health and well-being of all individuals but is especially critical in the elderly because immune responses generally decline with age (Miller, 1994; Chandra, 1995). The consequences of suboptimal immune responses are particularly detrimental in the elderly, who have an increased risk of infections as well as immune-mediated cancers, adverse hospital outcomes, and autoimmune diseases (Meydani and Blumberg, 1993).

As cells and tissues age, there is frequently an alteration in cell membrane components that mark the aged cells for elimination by macrophages. Laboratory studies have shown that red blood cells from vitamin E-deficient rats have more of the "senescent cell antigen" than red cells from vitamin E-sufficient rats, resulting in their increased phagocytosis and destruction by macrophages. Vitamin E deficiency caused other changes, including oxidative damage in red cells, which also suggested the premature aging of these cells (Kay et al., 1986).

The cell-mediated immune responses involving T-lymphocyte functions (cytotoxicity, interleukin-2 production, proliferation) are the most sensitive to the age-related decline in immune responses (Ernst, 1995). As a consequence, DTH responses to skin test antigens are significantly diminished in the elderly, and can often result in complete loss of response to antigen challenge (anergy) in the most immunosuppressed (Christou, 1990). Clinical studies have shown that DTH can be used as a predictor of morbidity and mortality in the elderly; for example, elderly individuals with anergy had twice the risk of death from all causes as elderly individuals who responded to the antigens (Wayne et al., 1990). Moreover, in hospitalized elderly patients who had undergone surgery for any reason, anergy was associated with a greater than 10-fold increased risk of mortality and a 5-fold increased risk of sepsis (Christou, 1990). Thus, if micronutrient supplements could improve DTH responses in the elderly, the health effects could be very great (Goodwin, 1995).

In fact, in a recent well-controlled study, intake of a one-a-day type multivitamin–mineral supplement for 12 months significantly enhanced DTH in healthy elderly subjects (Bogden et al., 1994). Because the multivitamin included β-carotene, vitamin E, vitamin C, and all other essential vitamins as well as several minerals, it is not possible to determine whether the antioxidants or the other components in the multivitamin supplement were responsible for the improved DTH responses. There are, however, several studies discussed later that have examined the effects of individual antioxidants on DTH responses in the elderly.

There is evidence that a supplement containing antioxidants can improve the immune cell profile of aged hospitalized patients. Elderly patients who were hospitalized for a least 2 months following a stroke were given a supplement of 8000 IU of vitamin A, 50 IU of vitamin E, and 100 mg of vitamin C for 28 days. Supplementation resulted in an increase in total T-lymphocyte numbers and T-helper cell markers as well as enhanced lymphocyte proliferation. Improvement in the immunosuppressed state of long-term, hospitalized elderly patients may decrease the prolonged morbidity often associated with respiratory infections seen in this population (Penn et al., 1991).

Recent data suggest that NK cell capacity to kill tumor cells is reduced in the elderly (Santos et al., 1996). NK and other cells of the immune system are responsible for internal surveillance, identification, and destruction of potential cancer cells. It is hypothesized that the increase in cancer risk with advanced age is a consequence of age-related diminution in immune responses, such as NK, which are necessary for cancer prevention.

The capacity to survive pneumonia and influenza is significantly reduced in the elderly ("Increasing influenza," 1995). Of the 20,000 pneumonia- and influenza-associated deaths reported in 1991 in the United States, 90% were in persons age 65 and older. There is a vaccine for bacterial pneumonia; however, it is only 56% effective in preventing pneumococcal pneumonia, the most common cause of bacterial pneumonia. Currently only 28% of the elderly receive pneumococcal vaccination; 52% receive influenza vaccination. Thus it is especially important to determine whether antioxidants and/or other nutrients can improve immune responses in this at-risk population, because another major consequence of reduced immune responses in the elderly is decreased effectiveness of vaccination (which can be predicted by reduced antibody titer following inoculation). Poor immune responses to vaccines such as the pneumococcal, hepatitis, and/or flu vaccine can increase the risk of morbidity and mortality, especially in the frail elderly. In theory, enhancement in DTH should translate into improved vaccination responses. Several of the intervention studies discussed here examined whether there is, in fact, concordance between enhanced DTH and improved vaccination responses with improved antioxidant status.

1. *Effects of Antioxidants on Immune Functions in the Elderly*

The well documented decline in immune responses with aging and the evidence of increased free radical damage in the elderly has led to a number of clinical intervention trials to determine whether antioxidant supplementation can beneficially affect immune functions in the elderly (Table I). The studies reviewed here involved healthy, free-living elderly individuals who are sufficiently motivated to become participants in clinical trials.

TABLE I
Effects of Supplements on Immune Parameters from Intervention Studies Involving Healthy Elderly Adults

Reference	Substance	Duration (months)	Placebo controlled/ double blind	Number of study subjects	DTH	Vaccine titer	Mitogen	IL-2 receptor/ product	NK number	Lymphocyte Total	Lymphocyte Subsets
Bodgen et al. (1994)	Multivitamin	12	Yes	56	✓	—	—	—	—	—	—
Herraiz et al. (1996)	BC (30 mg)	2	Yes	32	✓	—	—	—	—	—	—
Meydani et al. (1995)	BC (90 mg)	.75	Yes	NS	✓	—	NE	NE	—	—	—
Watson et al. (1991)	BC (45 or 60 mg)	2	No	20	—	—	—	✓	✓	NE	NE
Santos et al. (1996)	BC (50 mg every other day)	~144	Yes	59	—	✓	—	NE	NE/✓ killing	NE	✓
Chandra (1992)	Multi with 16 mg BC	12	Yes	96	—	✓	—	—	—	—	—
Meydani et al. (1990)	Vitamin E (800 IU)	1	Yes	32	✓	—	✓	✓	—	—	—
Meydani et al. (1994a)	Vitamin E (60, 200, or 800 IU)	6	Yes	80	✓	✓	✓	—	—	NE	NE
Pike and Chandra (1995)	Multi with 45 IU vit E, 90 mg vit C	12	Yes	47	—	—	—	—	✓	No decline	—

Key: Check mark, significant positive effect in at least one assay per study in supplemented group; dash, not determined; NE, no effect seen; NS, not stated by authors.

a. β-Carotene. Nutritionists have traditionally viewed β-carotene solely as a source of vitamin A activity in the diet. However, several studies have shown that carotenoids can enhance immune functions independently of any provitamin A activity (Bendich, 1994). The mechanisms of immunoenhancement may include the antioxidant- and singlet oxygen-quenching capacities of β-carotene as well as a number of other carotenoids.

In laboratory studies, comparisons between β-carotene and the nonvitamin A carotenoid, canthaxanthin have shown that both carotenoids enhanced T- and B-lymphocyte proliferative responses to mitogens; increased cytotoxic T-cell and macrophage tumor-killing activity, and stimulated the secretion of tumor necrosis factor-α while lowering the tumor burden (reviewed in Bendich, 1991).

In a clinical study, vegetarians were found to have similar serum levels of all of the vitamins compared to a matched nonvegetarian population. However, serum β-carotene levels were two times higher in the vegetarian group. NK from the vegetarian group lysed twice the number of tumor cells as compared to NK from the nonvegetarian group, suggesting that β-carotene or other carotenoids may enhance NK functions independently of their provitamin A activity (Malter *et al.,* 1989). Kramer *et al.* (1995), in a preliminary report, found that carotenoid-rich vegetable consumption enhanced NK cell number as well as lymphocyte proliferation.

Because β-carotene is an excellent quencher of singlet oxygen, which is thought to be formed during photosensitivity reactions, and because exposure to ultraviolet (UV) rays in sunlight can cause immunosuppression as determined by DTH responses, there has been an interest in the potential for β-carotene to reduce UV-induced immunosuppression. *In vitro,* β-carotene prevented UV-induced reduction in human cell function (Schoen and Watson, 1988). In animal studies, β-carotene reduced UV-induced tumor formation (Tomita *et al.,* 1987). White *et al.* (1988) reported a significant decrease in circulating plasma carotenoids following exposure to UV light. Carotenoids containing nine or more conjugated double bonds can quench the high energy from UV light and block the subsequent formation of singlet oxygen. In a recent study in healthy males, β-carotene supplementation prevented UV-induced depression in DTH responses (Fuller *et al.,* 1992).

The immunosuppressive effects of exposure to UV light have been documented to decrease the activities of non-Langerhan's epidermal antigen-presenting cells (Baadsgaard and Wang, 1991) and to reduce interleukin production by T cells (Araneo *et al.,* 1989; Schwarz and Luger, 1989). UV exposure causes a significant reduction in circulating total lymphocytes and helper T lymphocytes, resulting in an inversion of the helper–suppressor ratio (Hersey et al., 1983). Furthermore, UV exposure decreases antioxidant enzyme levels in the skin while increasing the skin's lipid hydroperoxide levels (Shindo et al., 1993).

There have not been many investigations of the effects of UV light on immune responses in the elderly even though this is an important practical consideration; many of the elderly have increased exposure to sunlight in retirement years compared to when they were employed, and a significant number of retirees relocate to areas of greater sunlight exposure, such as Florida, California, and Arizona. Thus the increased UV exposure coupled with a declining immune response could further increase the risk of infection.

Herraiz et al. (1996) have examined the effect of β-carotene in healthy elderly subjects who were exposed to a controlled level of UV light. In a placebo-controlled, double-blind trial, 32 males about 65 years of age had baseline DTH responses determined and then were placed on a low-carotene diet, and all were given 1.5 mg of β-carotene per day (the U.S. average daily intake) for the duration of the study. The supplement (30 mg β-carotene) or the matched placebo was taken for 28 days prior to UV exposure and a pre-UV DTH response was determined. The third and final DTH response was determined following 12 exposures over 16 days to a total of 15 J/cm^2 of UV light; the supplement or placebo was taken during this time. The placebo group showed a 50% decline in serum β-carotene between baseline and pre-UV DTH responses, whereas the supplemented group had a 10-fold increase in serum β-carotene. There were no changes seen in the β-carotene-supplemented group's serum levels of the other carotenoids or vitamin E throughout the study.

Baseline and pre-UV DTH responses did not differ between groups. However, there was a significant decline in the number of DTH responses as well as total induration in the placebo group following exposure to UV compared to their pre-UV responses, which was not seen in the β-carotene-supplemented group. There was also a decline in DTH responses of the placebo group who were not exposed to UV light, such that the number of responses of the placebo group as a whole was significantly reduced compared to baseline values over the course of the study. In comparison, the β-carotene group not exposed to UV light showed a slight increase in DTH between the second and third determinations, and the supplemented group exposed to UV light had a smaller decline in DTH, resulting in a nonsignificant change in DTH over course of the study (Herraiz et al., 1996).

This study demonstrated that healthy elderly men exposed to relatively low levels of UV light for short periods of time each day (2–6 min/day) had measurable decreases in their DTH responses. Furthermore, β-carotene supplementation at 30 mg/day protected the elderly men in this study from UV-induced immunosuppression significantly better than 1.5 mg/day of β-carotene. It should be noted, however, that the 1.5 mg/day of β-carotene did afford some benefit because, in an earlier similar protocol, young men fed a β-carotene-depleted diet had significantly greater declines in DTH following UV exposure (Fuller *et al.*, 1992) than were seen in the placebo group in this study.

Meydani *et al.* (1995) reported preliminary results from a short-term, placebo-controlled, double-blind intervention study in which elderly women (average age was 70 years) were given 90 mg/day of β-carotene for 3 weeks and were not exposed to UV light. Serum β-carotene levels increased 10-fold in the supplemented group; there was also a significant 35% increase in DTH ($p<.05$). There were no effects on mitogen responses, interleukin-2 (IL-2) production, or total lymphocyte count and lymphocyte subsets; there was no report of effects on NK cell number or activity. Concomitantly, there was an increase in the antioxidant capacity of the plasma of the β-carotene-supplemented group (M. Meydani *et al.*, 1994).

There have been several studies to determine the effect of β-carotene supplementation on NK function in the elderly. In the first published report, healthy middle-aged individuals (average age 56 years) were supplemented for 2 months with either 0, 15, 30, 45, or 60 mg/day of β-carotene. The groups given 45 or 60 mg/day had increases in the percent of T helper and NK cells and a concomitant increase in immune cells expressing the IL-2 receptor (Watson *et al.*, 1991). IL-2, a cytokine secreted by activated T cells, is thought to bind to the IL-2 receptor on NK cells and result in proliferation of NK cells. In this short-term study, however, there was no determination of either IL-2 production or the capacity of the NK cells to kill tumor cells.

Recently, Santos *et al.* (1996) examined the effect of long-term β-carotene supplementation on NK cell number and function as well as production of the NK cell stimulatory cytokine, IL-2. The study population was a cohort from the ongoing Physicians Health Study, a placebo-controlled, double-blind intervention trial to determine the effects of aspirin and/or β-carotene (50 mg every other day for more than 10 years) on cancer and cardiovascular outcomes. The subjects in the cohort examined by Santos *et al.* were divided into two groups based upon age. The middle-aged (MA) group's average age was 57 and the elderly (E) were about 73 years old. These two groups were further divided into β-carotene (BC)-supplemented and matched placebo (P) groups; both groups took aspirin every other day for at least the last 5 years.

There was a 3-fold increase in β-carotene serum levels in the supplemented groups, regardless of age (Fotouhi et al., 1995). White blood cell β-carotene levels were increased 6-fold in the MA-BC group compared to the MA-P group but were only increased 3.5-fold in the E-BC compared to the E-P group. NK killing of tumor cells was significally diminished in the E-P compared to the MA-P group. However, the E-BC group had a significantly increased NK activity compared to the E-P group, which was increased to the level seen in the MA-BC group. In contrast to the study by Watson et al. (1991), there was no increase in NK cell number. No increase in IL-2 production was found, although there was a nonsignificant decrease in prostaglandin E_2 production in the β-carotene–supplemented group. These results are of particular importance because of their prospective nature. It will be of interest to determine whether there was a decreased risk of cancer in the (older) physicians who were supplemented with β-carotene.

In another recent placebo-controlled intervention trial, elderly subjects were given a multivitamin daily for 1 year that contained approximately eight times the standard level of intake of β-carotene (16 mg). The supplemented group had significantly fewer infections than the placebo group. Responses to vaccines were also improved in the supplemented group (Chandra, 1992). Further research is required to determine whether it was β-carotene in the supplement that was responsible for the beneficial effects.

b. Vitamin E. Laboratory animal studies have shown that the level of dietary vitamin E required to prevent overt signs of vitamin E deficiency, such as red blood cell hemolysis and testes degeneration, was insufficient to stimulate optimal immune response, even in young animals (Bendich, 1990). Thus, there was a strong rationale to examine the effects of supplemental levels of vitamin E in aged animals to determine whether their diminished immune responses could be improved.

Meydani et al. (1986) showed that the macrophages from aged mice secrete significantly greater levels of immunosuppressive prostaglandin E_2 than those from younger mice; supplementation with vitamin E lowered prostaglandin levels and enhanced *in vivo* cell-mediated immune responses of aged mice to the level seen in younger mice. Increased prostaglandin E_2 levels are correlated with decreased IL-2 production, which would also be immunosuppressive. Macrophages from vitamin E-deficient rats also secrete higher levels of prostaglandin E_2, which results from the activation of phospholipase A_2 and cyclooxygenase. Vitamin E in macrophage membranes inhibits the activation of these enzymes (Sakamoto et al., 1991).

One of the first studies to associate vitamin E status and immune function in the elderly was published in 1984. Chavance et al. (1984), in an epidemiological study, found a significant association between high plasma vitamin E levels and a lower number of infections in healthy adults over the age of 60. This report and the animal data led to several intervention studies.

Two placebo-controlled, double-blind studies indicate that vitamin E supplementation alone can significantly enhance DTH responses and/or T-cell subpopulations and proliferative responses as well as IL-2 activities in the elderly. In a carefully controlled study conducted in a metabolic ward where the daily diet contained approximately RDA levels of all nutrients, vitamin E supplementation (800 IU/day) for 1 month resulted in significantly increased DTH responses of healthy elderly (Meydani et al., 1990). Lymphocyte vitamin E levels increased over threefold with supplementation and were correlated with enhanced immune responses such as enhanced IL-2 production. The vitamin E-supplemented group showed no adverse effects (Meydani et al., 1994b); in fact, the beneficial effects included enhanced lymphocyte proliferation, decreased production of prostaglandin E_2, and decreased levels of serum lipid peroxides. As mentioned, the subjects in this study all consumed meals containing the recommended levels of all nutrients, including vitamin E, while they were residents of a metabolic ward. Thus, the diets and environments of the placebo and vitamin E groups were virtually equivalent. The authors concluded that "it is encouraging to note that a single nutrient supplement can enhance immune responsiveness in healthy elderly subjects consuming the recommended amounts of all nutrients."

Meydani et al. (1994a) extended their earlier findings and examined the effects of 6 months of supplementation with 60, 200, and 800 IU/day of vitamin E in a placebo-controlled, double-blind study in healthy, free-living elderly. In addition to DTH responses, they determined in vitro proliferation and ex vivo antibody titers to clinically relevant vaccines. DTH responses were significantly increased above placebo levels in all three supplemented groups; the greatest responses were seen in the 200-IU/day group. In vitro proliferative responses to the mitogen concanavalin A were the highest in the 800-IU/day group. Antibody titers to tetanus were unaffected by vitamin E supplementation; however, titers to hepatitis B vaccine were the highest in the 200-IU/day group. It is important to note that the current RDA for vitamin E for adults is 15 IU.

Recently, Pike and Chandra (1995) have shown an increase in the number of NK cells in 35 healthy elderly adults (69 years average age) who participated in a 1-year, placebo-controlled, double-blind study involving a multivitamin–mineral supplement that contained 45 IU of vitamin E and 90 mg of vitamin C but no β-carotene. NK activity was not reported. They also reported a significant decrease in CD4− T-helper cells in the placebo group, which resulted in a significant decline in the CD4–CD8 ratio over the 1-year period that was not seen in the supplemented group.

c. Vitamin C. There have been neither many nor recent studies of the effect of vitamin C on immune responses in the elderly. In one study, DTH was enhanced in an elderly population following injections of 500 mg/day of vitamin C (Kennes *et al.*, 1983). In another study, oral supplementation with vitamin C (2 g/day) in an elderly population enhanced *in vitro* lymphocyte proliferative responses but did not affect DTH (Delafuente *et al.*, 1986).

C. Cigarette Smoking

Approximately one-third of U.S. adult women and one-fourth of adult men smoke. The risk of lung cancer is approximately 15 times greater in smokers compared to nonsmokers (Byers *et al.*, 1987).

Cigarette smoke contains millions of free radicals per puff. In addition, several other harmful products in cigarette smoke can stimulate the formation of highly reactive molecules that further increase the free radical burden (Pryor and Stone, 1993). For example, Hoshino *et al.* (1990) demonstrated that smokers have significantly higher breath pentane levels than nonsmokers.

The micronutrient most affected by cigarette smoking appears to be the antioxidant vitamin C. The 1989 RDA mentions that smokers should consume at least 100 mg of vitamin C per day to reach the same serum vitamin C level as nonsmokers. Data from the second National Health and Nutrition Examination Survey (NHANES II) actually suggests that smokers require about 250 mg of vitamin C per day to reach the same blood levels as nonsmokers consuming the vitamin C RDA of 60 mg (Schectman *et al.*, 1991). Passive smokers have serum vitamin C levels that are intermediate between those of smokers and nonsmokers (Tribble *et al.*, 1993). Tobacco chewers have lower vitamin C levels than nonsmokers, and similar vitamin E levels (Giraud *et al.*, 1995). In addition to vitamin C, serum levels of vitamin E, folic acid, and β-carotene as well as lung vitamin E concentrations are significantly lower in smokers compared to nonsmokers (Bendich, 1994).

Cigarette smokers have depressed immune responses compared to nonsmokers (Holt, 1987; McSharry and Wilkinson, 1986; Johnson et al., 1990), which may in part be due to the overproduction of immunosuppressive free radicals by phagocytic neutrophils and macrophages in the lungs of smokers. The lung of the healthy nonsmoker contains very few neutrophils. However, there are increased numbers of neutrophils and macrophages in the lungs of smokers (Pacht et al., 1986).

In smokers, there is a constant activation of neutrophils and a consequent overproduction of reactive oxygen species. The activated oxygen radicals can then interact with the α_1-proteinase inhibitor in the lung and inactivate this protective enzyme (Theron and Anderson, 1985). Mobrahan et al. (1990) showed that supplementation of smokers with 30 mg of β-carotene for 4 weeks decreased the overactivity of phagocytes. Using a separate assay system, Clausen (1992) found that β-carotene supplementation in smokers reduced the oxidant stress-related chemiluminescence of their neutrophils.

Epidemiological studies have consistently found that smokers with the highest intake of β-carotene-rich foods have significant reduction in risk of cancers of the lung, mouth, esophagus, and stomach (Block, 1992). The average intake of β-carotene in the high-intake groups is about 6–10 mg/day. Van Antwerpen et al. (1995) examined the association between oxidant generation from neutrophils from smokers and serum β-carotene levels. They found that, compared to nonsmokers, smokers had significantly lower serum β-carotene levels (10.9 versus 14.4 mg/dl; $p<0.01$), significantly more neutrophils in peripheral blood (41%; $p<0.001$), and significantly greater oxidant generation (47%; $p<0.004$). The increase in oxidants was a result of increase in neutrophil number rather than increased oxidant production per neutrophil. Serum β-carotene was inversely correlated to neutrophil count and oxidative activity only in the smokers, suggesting that the reduction in β-carotene was due to its utilization as an antioxidant. In a placebo-controlled, double-blind study, smokers given 20 mg/day of β-carotene for 14 weeks had a significant reduction in abnormal cells in their sputum but no change in the level of DNA damage in circulating white blood cells (Richards et al., 1990; van Poppel et al., 1992, 1993). Abnormal sputum cells may be an early sign of precancerous oral, bronchial, or lung lesions, which are precursors of cancer. In a second study, β-carotene supplementation (30 mg/day for 2 months) reduced the progression of oral precancerous lesions in 70% of smokers while enhancing NK cell receptors and in vitro killing of tumor cells by NK cells (Garewal, 1992). Lymphocyte proliferation is also depressed in smokers and has been shown to be increased when smokers were given 20 mg/day of β-carotene for 14 weeks (van Poppel et al., 1993). Given the data from these short-term studies with β-carotene supplementation, the results of the Alpha-Tocopherol Beta Carotene Cancer Prevention Group study (1994) that β-carotene supplementation increased the risk of lung cancer in heavy smokers are surprising and inexplicable.

Hoshino et al. (1990) showed that supplementation with 800 IU of vitamin E for 2 weeks resulted in a significant decrease in breath pentane levels in smokers. Richards et al. (1990) found that circulating phagocytes of smokers produced high levels of free radicals. Administration of 900 IU/day of vitamin E for 6 weeks to these smokers significantly reduced the overproduction of oxidant radicals by circulating phagocytic cells.

Smokers have been shown to have significantly lower concentrations of vitamin E in their lungs than nonsmokers; however, even supplementation with 2400 IU/day (80 times the daily value) for 3 weeks failed to restore the lung vitamin E level to that found in nonsmokers (Pacht et al., 1986). Pacht et al. also reported that oral supplementation with vitamin E did not result in a reduction in the *in vitro* overproduction of superoxide anions from alveolar macrophages of smokers. Of significance, those smokers with the lowest lung vitamin E levels had the greatest killing of lung tissue cells. Thus, a decreased vitamin E concentration in the lungs of smokers may result in lung tissue destruction from the highly cytotoxic macrophages found there (Davis et al., 1988).

VII. Conclusions

Immune function is dependent upon a balance between the free radical and antioxidant status of the individual. In healthy adults, exposure to high levels of oxidants is associated with reduction in clinically relevant immune responses; exposure to low levels of dietary antioxidants also reduces immune responses such as DTH.

Antioxidant status becomes particularly critical in conditions that result in increased oxidative stress, as illustrated in RA, aging, and cigarette smoking. In all three examples, evidence of free radical damage to lipids and other cellular components have been demonstrated. At the same time, antioxidant status is reduced in arthritic patients and smokers compared to controls. Data suggest that elderly adults may require higher levels of antioxidants to reach the same blood levels seen in younger controls. It is well recognized that smokers require higher levels of vitamin C and β-carotene intake to reach blood levels found in nonsmokers.

Moreover, when antioxidant vitamins have been provided, there is a reduction in the indices of oxidative stress and a concomitant improvement in several measures of immune function. The data are particularly impressive from recent well-controlled studies in the healthy elderly. There appears to be a consistent improvement in DTH responses when antioxidant supplements are provided for relatively short periods of time, and the improvement appears to persist for the duration of the supplementation. The recent determination that vaccination antibody titers are also increased in the healthy elderly adds to the relevance of the simple, safe (Bendich, 1992; Bendich and Langseth, 1995), and inexpensive practice of modest supplementation. Because of the promising results in the elderly and the health care cost implication, studies of longer duration, with a greater number of subjects and repetition of the vaccination findings, are recommended. With regard to the inflammatory autoimmune disease, RA, the data are more preliminary and suggest a dose–response relationship in which vitamin E may afford modest reductions in pain. More well-controlled, large-scale clinical trials are needed to define the optimal dose–duration and mechanisms of action. With regard to cigarette smoking, it is accepted that this habit has many detrimental health effects, including the oxidant-related decreases in immune response. Although the best advice to give to smokers is cease smoking, this is unrealistic. About 10% of smokers who want to stop smoking have stopped for 1 year even when they are enrolled in smoking cessation programs. Thus, it remains necessary to determine the antioxidant requirement of smokers, and include immune responses as part of the outcome measures.

Although further research is warranted, there are sufficient data currently available to recommend that those at risk for oxidative stress and/or low antioxidant intake increase their intake of antioxidant vitamins. At the very least, health professionals should provide patients with information on dietary sources for meeting the current RDAs as well as information on the safety and efficacy of antioxidant vitamins. Health professionals should also discuss the results of clinical studies with patients so that they can make informed choices.

REFERENCES

Aho, K., Heliovaara, M., Maatela, J., Tuomi, T., and Palosuo, T. (1991). Rheumatoid factors antedating clinical rheumatoid arthritis. *J. Rheumatol.* **18**, 1282–1284.

Alexander, J. W. (1995). Specific nutrients and the immune response. *Nutrition* **11**, 229–232.
Alpha-Tocopherol Beta Carotene Cancer Prevention Group. (1994). The effect of vitamin E and beta carotene on the incidence of lung cancer and other cancers in male smokers. *New Engl. J. Med.* **330**, 1029–1035.
Anderson, R., Smit, M. J., Joone, G. K., and Van Staden, A. M. (1990). Vitamin C and cellular immune functions. *Ann. N.Y. Acad. Sci.* **587**, 34–48.
Araneo, B. A., Dowell, T., Moon, H. B., and Daynes, R. A. (1989). Regulation of murine lymphokine production in vivo: UV radiation exposure depresses IL2 and enhances IL4 production by T cells through an IL1 dependent mechanism. *J. Immunol.* **143**, 1737–1744.
Azzini, M., Girelli, D., Olivieri, O., Guarini, P., Stanzial, A. M., et al. (1995). Fatty acids and antioxidant micronutrients in psoriatic arthritis. *J. Rheumatol.* **22**, 103–108.
Baadsbaard, O., and Wang, B. S. (1991). Immune regulation in allergic and irritant skin reactions. *Int. J. Dermatol.* **30**, 161–171.
Baker, K. R., and Meydani, M. (1994). Beta-carotene in immunity and cancer. *J. Optim. Nutr.* **3**, 39–50.
Beck, M. A., Kolbeck, P. C., Rohr, L. H., Shi, Q., Morris, V. C., and Levander, O. A. (1994a). Vitamin E deficiency intensifies the myocardial injury of coxsakievirus B3 infection of mice. *J. Nutr.* **124**, 345–358.
Beck, M. A., Kolbeck, P. C., Rohr, L. H., Shi, W., Morris, V. C., and Levander, O. A. (1994b). Increased virulence of a human enterovirus (coxsackievirus B3) in selenium-deficient mice. *J. Infect. Dis.* **170**, 351–357.
Beck, M. A., Shi, Q., Morris, V. C., and Levander, O. A. (1995). Rapid genomic evolution of a non-virulent coxsackievirus B3 in selenium-deficient mice results in selection of identical virulent isolates. *Nature Med.* **1**, 433–436.
Beckman, J. S., and Crow, J. P. (1993). Pathological implications of nitric oxide, superoxide and peroxynitrite formation. *Biochem. Soc. Trans.* **21**, 330–334.
Bendich, A. (1990). Antioxidant micronutrients and immune responses. *In* "Micronutrients and Immune Functions" (A. Bendich and C. E. Butterworth, eds.), Vol. 87, pp. 168–180. New York Academy of Sciences, New York.
Bendich, A. (1991). Carotenoids and immunity. *Clin. Appl. Nutr. Immune Function* **1**, 45–51.
Bendich, A. (1993). Clinical importance of beta carotene. *Clin. Appl Nutr.* **1**, 14–22.
Bendich, A. (1994). Recent advances in clinical research involving carotenoids. *Pure Appl. Chem.* **66**, 1017–1024.
Bendich, A. (1995). Criteria for determining Recommended Dietary Allowances for healthy older adults. *Nutr. Rev.* **53**, S105–S110.
Bendich, A., and Langseth L. (1995). The health effects of vitamin C supplementation: a review. *J. Am. Coll. Nutr.* **14(2)**:124–36.
Bendich, A., and Cohen, M. (1996). Vitamin E, rheumatoid arthritis and other arthritic disorders. *J. Nutr. Immunol.*, in press.
Bendich, A., Burton, G. W., Machlin, L. J., Scandurra, O., and Wayner, D. D. M. (1986). The antioxidant role of vitamin C. *Adv. Free Radical Biol. Med.* **2**, 419–444.
Block, G. (1992). The data support a role for antioxidants in reducing cancer risk. *Nutr. Rev.* **50**, 207–213.
Bogden, J. D., Bendich, A., Kemp, F. W., Bruening, K. S., Shurnick, J. H., et al. (1994). Daily micronutrient supplements enhance delayed-hypersensitivity skin test responses in older people. *Am. J. Clin. Nutr.* **60**, 437–447.
Burton, G. W., Joyce, A., and Ingold, K. U. (1983). Is vitamin E the only lipid-soluble,

chain-breaking antioxidant in human blood plasma and erythrocyte membranes? *Arch. Biochem. Biophys.* **221**, 281–290.

Byers, T. E., Graham, S., Haughey, B. P., Marshall, J. R., and Swanson, M. K. (1987). Diet and lung cancer risk: Findings from the Western New York Diet Study. *Am. J. Epidemiol.* **125**, 351–363.

Chandra, R. K. (1992). Effect of vitamin and trace-element supplementation on immune responses and infection in elderly subjects. *Lancet* **340**, 1124–1127.

Chandra, R. (1995). Nutrition and immunity in the elderly: Clinical significance. *Nutr. Rev.* **53**, S80–S83.

Chandra, R. K., and Kumari, S. (1994). Effects of nutrition on the immune system. *Nutrition* **10**, 207–210.

Chavance, M., Brubacher, G., Herbeth, B., Vernhes, G., Mikstacki, T., et al. (1984). Immunlogical and nutritional status among the elderly. *Topics Aging Res. Eur.* **1**, 231–237.

Chew, B. P. (1995). Antioxidant vitamins affect food animal immunity and health. *J. Nutr.* **125**, 1804S–1805S.

Christou, N. (1990). Perioperative nutrition support: Immunologic defeats. *JPEN J. Parenter. Enteral Nutr.* **14**, 186S.

Clausen, J. (1992). The influence of antioxidants on the enhanced respiratory burst reaction in smokers. *Ann. N.Y. Acad. Sci.* **669**, 337–341.

Coltro, J. R., Zarling, E. J., and Skosey, J. L. (1987). Correlation of pentane, a marker of oxygen radical action, with therapeutic response of patients with rheumatoid arthritis (RA) and systemic lupus erythematisus (SLE). *Clin. Res.* **35**, 894A.

Davis, W. B., Pacht, E. R., Spatafora, M., and Martin, W. J. (1988). Enhanced cytotoxic potential of alveolar macrophages from cigarette smokers. *J. Lab. Clin. Med.* **111**, 293–298.

Delafuente, J. C., Prendergast, J. M., and Modigh, A. (1986). Immunologic modulation by vitamin C in the elderly. *Int. J. Immunopharmacol.* **8**, 205–211.

Ernst, D. N. (1995). Aging and lymphokine gene expression by T cell subsets. *Nutr. Rev.* **53**, S18–S26.

Fotouhi, N., Meydani, M., Santos, M., Meydani, S. N., Hennekens, C. H., and Gaziano, J. M. (1995). The effect of long term B-carotene (B-C) supplementation on carotenoids and tocopherol concentration in plasma, RBC, and peripheral blood mononuclear cells (PBMC). *FASEB J.* **9**, A170.

Frei, B. (1991). Ascorbic acid protects lipids in human plasma and low-density lipoprotein against oxidative damage. *Am. J. Clin. Nutr.* **54**, 1113S–1118S.

Fuller, C. J., Faulkner, H., Bendich, A., Parker, R. S., and Roe, D. A. (1992). Effect of beta-carotene supplementation on photosuppression of delayed-type hypersensitivity in normal young men. *Am. J. Clin. Nutr.* **56**, 684–690.

Garewal, H. S. (1992). Potential role of beta-carotene and antioxidant vitamins in the prevention of oral cancer. *Ann. N.Y. Acad. Sci.* **669**, 261–267.

Giraud, D. W., Martin, H. D., and Driskell, J. A. (1995). Erythrocyte and plasma B-6 vitamin concentrations of long-term tobacco smokers, chewers, and nonusers. *Am. J. Clin. Nutr.* **62**, 104–109.

Goodwin, J. S. (1995). Decreased immunity and increased morbidity in the elderly. *Nutr. Rev.* **53**, S41–S46.

Halliwell, B. (1995). Oxygen radicals, nitric oxide and human inflammatory joint disease. *Ann. Rheum. Dis.* **54**, 505–510.

Harris, E. D. (1985). Pathogenesis of rheumatoid arthritis. *In* "Textbook of Rheumatology" (W. N. Kelley, E. D. Harris, S. Ruddy, and C. B. Sledge, eds.), pp. 886–915. W. B. Saunders Company, Toronto.

Heliovaara, M., Knekt, P., Aho, K., Aaran, R. K., and Alfthan, G., (1994). Serum antioxidants and risk of rheumatoid arthritis. *Ann. Rheum. Dis.* **53,** 51–53.

Herraiz, L. A., Hsieh, W. C., Parker, R. S., Swanson, J. E., Bendich, A., et al. (1996). Effect of B-carotene supplementation on photosuppression of delayed-type hypersensitivity in healthy older men. In press.

Hersey, P., Prendergast, D., and Edwards, A. (1983). Effects of cigarette smoking on the immune system: Follow-up studies in normal subjects after cessation of smoking. *Med. J. Aust.* **2,** 425–429.

Holt, P. (1987). Immune and inflammatory function in cigarette smokers. *Thorax* **42,** 241–249.

Hoshino, E., Shariff, R., Van Gossum, A., Allard, J. P., Pichard, C., et al. (1990). Vitamin E suppresses increased lipid peroxidation in cigarette smokers. *JPEN J. Parenter. Enteral Nutr.* **14,** 300–305.

Humad, S., Zarling, E. D., Clapper, M., and Skosey, J. L. (1988). Breath pentane excretion as a marker of disease activity in rheumatoid arthritis. *Free Radical Res. Commun.* **5,** 101–106.

Increasing influenza vaccination rates for Medicare beneficiaries—Montana and Wyoming, 1994. (1995). *MMWR* **44,** 744–746.

Jacob, R. A., Kelley, D. S., Pianalto, F. S., Swendseid, M. E., Henning, S. M., et al. (1991). Immunocompetence and oxidant defense during ascorbate depletion of healthy men. *Am. J. Clin. Nutr.* **54,** 1302s–1309s.

Johnson, J. D., Houchens, D. P., Kluwe, W. M., Craig, D. K., and Fisher, G. L. (1990). Effects of mainstream and environmental tobacco smoke on the immune system in animals and humans: A review. *CRC Crit. Rev. Toxicol.* **134,** 356–361.

Kaur, H., and Halliwell, B. (1994). Evidence for nitric oxide-mediated oxidative damage in chronic inflammation: Nitrotyrosine in serum and synovial fluid from rheumatoid patients. *FEBS Lett.* **350,** 9–12.

Kay, M. M., Bosman, G. J., Shapiro, S. S., Bendich, A., and Bassel, P. S. (1986). Oxidation as a possible mechanism of cellular aging: Vitamin E deficiency causes premature aging and IgG binding to erythrocytes. *Proc. Natl. Acad. Sci. U.S.A.* **83,** 2462–2467.

Kennes, B., Dumont, I., Brochee, D., Hubert, C., and Neve, P. (1983). Effect of vitamin C supplements on cell-mediated immunity in old people. *Gerontology* **29,** 305–310.

Kolarz, G., Scherak, O., El Shohoumi, M., and Blankenhorn, G. (1990). High-dose vitamin E in rheumatoid arthritis. *Akt. Rheumatol.* **15,** 233–237.

Kowsari, B., Finnie, S. K., Carter, R. L., Love, J., Katz, S., et al. (1983). Assessment of the diet of patients with rheumatoid arthritis and osteoarthritis. *J. Am. Diet. Assoc.* **82,** 657–659.

Kramer, T. R., Schoene, N., Douglass, L. W., Judd, F. T., Ballard-Barbash, R., et al. (1991). Increased vitamin E intake restores fish oil induced suppressed blastogenesis of mitogen stimulated T-lymphocytes. *Am. J. Clin. Nutr.* **54,** 896–902.

Kramer, T. R., Burri, B. J., and Neidlinger, T. R. (1995). Carotenoid–flavonoid modulated immune responses in women. *FASEB J.* **9,** A170.

Krinsky, N. I. (1989). Carotenoids in medicine. *In* "Chemistry and Biology" (N. I. Krinsky, M. M. Mathews-Roth, and R. F. Taylor, eds.), pp. 279–291. Plenum Press, New York.

Langley, G. B., Sheppeard, H., and Wigley, R. D. (1983). Placebo therapy in rheumatoid arthritis. *Clin. Exp. Rheumatol.* **1,** 17–21.

Lunec, J., and Hill, C. (1984). Some immunological consequences of free radical-production in rheumatoid arthritis. *In* "Oxygen Radicals in Chemistry and Biology" (W. Bors, M. Saran, and D. Tait, eds.), pp. 939–945. de Gruyter, Berlin.

Machlin, L. J., and Bendich, A. (1987). Free radical damage: Antioxidant defenses. *FASEB J* **1,** 441–445.

Malter, M., Schriever, G., and Eilber, U. (1989). Natural killer cells, vitamins, and other blood components of vegetarian and omnivorous men. *Nutr. Can.* **12,** 271–278.

McSharry, C., and Wilkinson, P. C. (1986). Cigarette smoking and the antibody response to inhaled antigens. *Immunol. Today* **7,** 98.

Merry, P., Winyard, P. G., Morris, C. J., Grootveld, M., and Blake, R. D. (1989). Oxygen free radicals, inflammation, and synovitis: The current status. *Ann. Rheum. Dis.* **48,** 864–870.

Meydani, M., Martin, A., Ribaya-Mercado, J. D., Gong, J., Blumberg, J. B., and Russell, R. M. (1994). B-carotene supplementation increases antioxidant capacity of plasma in older women. *J. Nutr.* **124,** 2397–2403.

Meydani, S. N., and Blumberg, J. B. (1993). Vitamin E and the immune response. In "Nutrient Modulation of the Immune Response" (J. Cunningham-Rundles, ed.), pp. 223–238. Marcel Dekker, New York

Meydani, S. N., Meydani, M., Verdon, C. P., Blumberg, J. P., and Hayes, K. C. (1986). Vitamin E supplementation suppresses prostaglandin E synthesis and enhances the immune response of aged mice. *Mech. Ageing Dev.* **34,** 191–201.

Meydani, S. N., Yogeeswaran, G., Liu, S., Baskar, S., and Meydani, M. (1988). Fish oil and tocopherol induced changes in natural killer cell mediated cytotoxicity and PGE2 synthesis in young and old mice. *J. Nutr.* **118,** 1245–1252.

Meydani, S. N., Barklund, M. P., Liu, S., Miller, R. A., Cannon, J. G., et al. (1990). Vitamin E supplementation enhances cell-mediated immunity in healthy elderly subjects. *Am. J. Clin. Nutr.* **52,** 557–563.

Meydani, S. N., Leka, L., and Loszewski, R. (1994a). Long-term vitamin E supplementation enhances immune response in healthy elderly. *FASEB J.* **8,** A272.

Meydani, S. N., Meydani, M., Rall, L. C., Morrow, F., and Blumberg, J. B. (1994b). Assessment of the safety of high-dose, short-term supplementation with vitamin E in healthy older adults. *Am. J. Clin. Nutr.* **60,** 704–709.

Meydani, S. N., Santos, M. S., Ribaya-Mercado, J. D., Leka, L., Han, S. N., and Russell, R. M. 1995). Effect of B-carotene (B-C) on the immune response of elderly women. *FASEB J.* **9,** A170.

Miller, R. A. (1994). Aging and immune function: Cellular and biochemical analyses. *Exp. Gerontol.* **29,** 21–35.

Mobarhan, S., Bowen, P., Andersen, B., Evans, M., Stacewicz-Sapuntzakis, M., et al. (1990). Effects of B-carotene repletion on b-carotene absorption, lipid peroxidation, and neutrophil superoxide formation in young men. *Nutr. Cancer* **14,** 195–206.

Nath, J., Powledge, A., and Wright, D. (1988). Studies of signal transduction in the respiratory burst-associated stimulation of fMet-Leu-Phe-induced tubulin tyrosinolation and phorbol 12-myristate 13-acetate-induced posttranslational incorporation of tyrosine into multiple proteins in activated neutrophils and HN 60 cells. *J. Biol. Chem.* **264,** 848–855.

Pacht, E. R., Kasek, H., Mohammad, J. R., Cromwell, D. G., and Davis, W. B. (1986). Deficiency of vitamin E in the alveolar fluid of cigarette smokers: Influence on alveolar macrophage cytotoxicity. *J. Clin. Invest.* **77,** 789–796.

Panayi, G. S., Lanchbury, J. S., and Kingsley, G. H. (1992). The importance of the T-cell in initiating and maintaining the chronic synovitis of rheumatoid arthritis. *Arthitis Rheum.* **35,** 729–735.

Penn, N. D., Purkins, L., Kelleher, J., Heatley, R. V., Mascie-Taylor, B. H., and Belfield, P. W. (1991). The effect of dietary vitamin supplementation with vitamins A, C and

E on cell-mediated immune function in elderly long-stay patients: A randomized controlled trial. *Age Aging* **20,** 169–174.

Pike, J., and Chandra, R. K. (1995). Effect of vitamin and trace element supplementation on immune indices in healthy elderly. *Int. J. Vitam. Nutr. Res.* **65,** 117–120.

Pryor, W. A., and Stone, K. (1993). Oxidants in cigarette smoke. *Ann. N.Y. Acad. Sci.* **686,** 12–28.

Richards, G. A., Theron, A. J., Van Rensburg, C. E. J., Van Rensburg, A. J., Van Der Merwe, C. A., et al. (1990). Investigation of the effects of oral administration of vitamin E and beta-carotene on the chemiluminescence responses and the frequency of sister chromatid exchanges in circulating leukocytes from cigarette smokers. *Am. Rev. Respir. Dis.* **142,** 648–654.

Sakamoto, W., Fujie, K., Nishihira, J., Mino, M., Morita, I. (1991). Inhibition of PGE2 production in macrophages from vitamin E-treated rats. *Prostaglandins Leukot. Essent. Fatty Acids.* **44,** 89–92.

Santos, M. S., Meydani, S. N., Leka, L., Wu, D., Fotouhi, N., et al. (1996). Elderly male natural killer cell activity is enhanced by B-carotene supplementation. In press.

Schectman, G., Byrd, J. C., and Hoffmann, R. (1991). Ascorbic acid requirements for smokers: Analysis of population survey. *Am. J. Clin. Nutr.* **53,** 1466–1470.

Schoen, D. J., and Watson, R. R. (1988). Prevention of UV irradiation induced suppression of monocyte functions by retinoids and carotenoids in vitro. *J. Photochem. Photobiol.* **48,** 659–663.

Schwarz, T., and Luger, T. A. (1989). Effects of UV irradiation on epidermal cell cytokine production. *J. Photochem. Photobiol.* **4,** 1–13.

Shindo, Y., Witt, E., and Packer, L. (1993). Antioxidant defense mechanisms in murine epidermis and dermis and their responses to ultraviolet light. *J. Invest. Dermatol.* **100,** 260–265.

Tappel, A. L., and Summerfield, F. W. (1984). Effects of dietary polyunsaturated fats and vitamin E on aging and peroxidative damage to DNA. *Arch. Biochem. Biophys.* **233,** 408–416.

Tarp, U. (1994). Selenium and the selenium-dependent glutathione peroxidase in rheumatoid arthritis. *Dan. Med. Bull.* **41,** 264–74.

Theron, A., and Anderson, R. (1985). Investigation of the protective effects of the antioxidants ascorbate, cysteine, and dapsone on the phagocyte-mediated oxidative inactivation of human alpha-1-protease inhibitor in vitro. *Am. Rev. Respir. Dis.* **132,** 1049–1054.

Thurnham, D. I., Situnayake, R. D., Koottathepis, S., McConkey, B., and Davis, M. (1987). Antioxidant status measured by the trap assay in rheumatoid arthritis. *In* "Free Radicals, Oxidant Stress and Drug Action" (C. Rice Evans, ed.), pp. 169–189. Richelieu Press, London.

Tomita, Y., Himeno, K., Nomota, K., Endo, H., and Hirohata, T. (1987). Augmentation of tumor immunity against syngeneic tumors in mice by beta carotene. *JNCI* **78,** 679–681.

Tribble, D. L., Giuliano, L. J., and Fortmann, S. P. (1993). Reduced plasma ascorbic acid concentrations in nonsmokers regularly exposed to environmental tobacco smoke. *Am. J. Clin. Nutr.* **58,** 886–890.

Tulleken, J. E., Limburg, P. C., Muskiet, F. A. J., and van Rijswijk, M. H. (1990). Vitamin E status during dietary fish oil supplementation in rheumatoid arthritis. *Arthritis Rheum.* **33,** 1416–1419.

Van Antwerpen, V. L., Theron, A. J., Richards, G. A., Van der Merwe, C. A., Viljoen, E., et al. (1995). Plasma levels of beta-carotene are inversely correlated with circulating neutrophil counts in young male cigarette smokers. *Inflammation* **19**, 405–414.

van Poppel, G., Kok, F. J., and Hermus, R. J. J. (1992). Beta-carotene supplementation in smokers reduces the frequency of micronuclei in sputum. *Br. J. Cancer* **66**, 1164–1168.

van Poppel, G., Spanhaak, S., and Ockhuizen, T. (1993). Effect of beta-carotene on immunological indexes in healthy male smokers. *Am. J. Clin. Nutr.* **57**, 402–407.

Watson, R. R., Prabhala, R. H.,. Plezia, P. M., and Alberts, D. S. (1991). Effect of beta-carotene on lymphocyte subpopulations in elderly humans: Evidence for a dose response relationship. *Am. J. Clin. Nutr.* **53**, 90–94.

Wayne, S. J., Rhyne, R. L., Garry, P. J., and Goodwin, J. S. (1990). Cell-mediated immunity as a predictor of morbidity and mortality in subjects over 60. *J. Gerontol. Med. Sci.* **45**, M45–M48.

White, W. S., Kim, C., Kalkwarf, H. J., Bustos, P., and Roe, D. A. (1988). Ultraviolet light-induced reductions in plasma carotenoid levels. *Am. J. Clin. Nutr.* **47**, 879–883.

Ziegler, R. G., and Subar, A. F. (1991). Vegetables, fruits and carotenoids and the risk of cancer. *In* "Micronutrients in Health and in Disease Prevention" (A. Bendich and C. E. Butterworth, eds.), pp. 97–126. Marcel Dekker, New York.

Zvaifler, N. J. (1988). New perspectives on the pathogenesis of rheumatoid arthritis. *Am. J. Med.* **85**, 12–17.

Cytokine Regulation of Bone Cell Differentiation

MELISSA ALSINA,* THERESA A. GUISE,† AND G. DAVID ROODMAN*,‡

Department of Medicine, Divisions of *Hematology and †Endocrinology, University of Texas Health Science Center at San Antonio, and ‡Audie Murphy Veterans Administration Hospital, San Antonio, Texas 78284

I. Introduction
II. Osteoblasts
   A. General Characteristics of the Osteoblast
   B. *In Vitro* Systems to Study the Osteoblast
   C. Factors Involved in Proliferation and Differentiation
III. Osteoclasts
   A. General Characteristics of the Osteoclast
   B. Systemic Hormones That Affect Osteoclast Function and Formation
   C. Autocrine–Paracrine Factors with Osteoclast-Stimulatory Activity
   D. Local Inhibitory Factors
IV. Summary
   References

I. Introduction

The activity of bone cells—osteoblasts and osteoclasts—is under control of both systemic hormones and cytokines generated in the bone microenvironment. These factors may act directly on osteoblasts, osteoclasts, or their precursors, or may act indirectly through an intermediary cell to control the formation, differentiation, and function of bone cells. In this article, we review the effects of these factors on bone cell differentiation and function.

II. Osteoblasts

Bone formation is the result of a series of events that involve proliferation of early mesenchymal cells, differentiation into osteoblast precursor cells, maturation of osteoblasts, formation of matrix, and, finally, mineralization. The major processes leading to bone formation include recruitment and replication of mesenchymal precursors of osteoblasts that are found in the periosteum and in the bone marrow adjacent to endosteal surfaces, their differentiation from preosteoblast to osteoblast, and regulation of their activity.

A. GENERAL CHARACTERISTICS OF THE OSTEOBLAST

The osteoblast produces a wide variety of proteins that compromise the bone matrix, including type I collagen, osteocalcin, matrix Gla protein, osteonectin, alkaline phosphatase, fibronectin, thrombospondin, proteoglycans, osteopontin, and bone sialoproteins. The osteoblast also secretes growth factors such as transforming growth factor-β (TGF-β), bone morphogenetic proteins (BMPs), insulin-like growth factor-I and -II (IGF-I, -II), platelet-derived growth factor (PDGF), and heparin-binding fibroblast growth factor (FGF), which are stored in bone matrix (Robey, 1989). These growth factors stimulate proliferation and differentiation of bone cells. Osteoblasts mineralize newly formed bone matrix, their presence appears to be required for osteoclastic bone resorption in rodents, and they enhance the bone-resorbing capacity of human osteoclasts. The osteoblast family also includes osteocytes and bone-lining cells. The function of these cells is not well understood, but they may play an intricate role in bone remodeling.

B. *In Vitro* SYSTEMS TO STUDY THE OSTEOBLAST

The development of experimental models used to study the osteoblast phenotype and bone formation has led to a better understanding of the function of cells of the osteoblast lineage. *In vitro* culture systems utilizing freshly dispersed bone cells, osteosarcoma cell lines, and nontransformed osteoblast-like cell lines derived from normal calvarial cells have been used to study differentiation of these cells in response to different osteotropic hormones. Unfortunately, these cell lines are not useful for determining lineage, and the variability in hormone responsiveness among cell lines makes it difficult to extrapolate such results to the *in vivo* situation (Mundy, 1995a,b).

Recently, an *in vitro* system of prolonged primary cultures of isolated fetal rat calvarial osteoblasts with β-glycerophosphate and ascorbic acid has been developed in which bone cells form clusters over several weeks and eventually become surrounded by woven bone (Bellows *et al.*, 1986). There is an initial stage of bone cell proliferation and extracellular matrix (ECM) biosynthesis followed by bone ECM development, maturation, and organization. The last phase is ECM mineralization. A developmental cascade of gene expression can be demonstrated in such cultures of fetal rat calvarial cells grown over several weeks. Specifically, during the proliferative phase, histone (reflecting DNA synthesis), *c-fos, c-jun,* and type I collagen gene expression are maximal. Histone, *c-fos,* and *c-jun* genes encode proteins that support proliferation of osteoblast precursors. Additionally, several genes associated with formation of the ECM, such as type I collagen, fibronectin, and TGF-β, are actively expressed during this proliferative phase of osteoblast development and then gradually down regulated during the subsequent stages of osteoblast differentiation. During the matrix maturation phase, when osteoblasts are postmitotic, alkaline phosphatase is expressed. The mineralization phase is characterized by maximal expression of the osteopontin, osteocalcin, and bone sialoprotein genes along with the accumulation of calcium (Stein and Lian, 1993). Thus, phenotypic markers of the differentiated osteoblast include alkaline phosphatase, osteopontin, and osteocalcin.

C. Factors Involved in Proliferation and Differentiation

1. *Systemic Factors*

a. Parathyroid Hormone. Parathyroid hormone (PTH) is a peptide produced by the parathyroid glands that is important in the maintenance of normal calcium homeostasis due to its stimulatory effects on osteoclastic bone resorption, renal tubular reabsorption of calcium, and production of 1α-hydroxylase activity in the kidney. When administered as a pharmacological agent, PTH has complex and different effects on bone depending on whether it is administered intermittently or continuously. Continuous infusion of PTH results in increased bone resorption. In contrast, administration of low doses of PTH in an intermittent fashion stimulates bone formation (Dempster *et al.,* 1993).

The primary target cell for PTH is the osteoblast (Rodan and Martin, 1981). Isolated osteoclasts do not resorb bone in response to PTH and only do so when primary cultures of osteoblasts or osteoblast-like cell lines are added (Chambers *et al.,* 1985; McSheehy and Chambers, 1986). Recent studies have described a distinct PTH target cell that is located among clusters of differentiated osteoblasts that exhibited structural features of a large soma and long cytoplasmic processes. Since these cells stained positively for alkaline phosphatase and displayed ultrastructural features of the differentiated osteoblast, they appear to be in the osteoblast lineage (Rouleau *et al.,* 1988, 1990). More recently, PTH receptors have been demonstrated in osteoclasts and osteoclast precursors (Teti *et al.,* 1991; Hakeda *et al.,* 1989) and, in fact, PTH has been shown to inhibit bone resorption in isolated osteoclasts

(Teti et al., 1991). Thus it appears that both osteoblasts and osteoclasts express functional PTH receptors and that PTH may have dual actions on bone: an indirect stimulatory effect mediated by osteoblasts and a direct inhibitory effect on the osteoclast (Dempster et al., 1993).

In vitro, PTH inhibits collagen synthesis, alkaline phosphatase activity, and osteocalcin synthesis in cultured osteoblastic cells (Dempster et al., 1993). However, PTH stimulates replication of osteoblasts at low doses in cultures of clonal osteosarcoma cells (Martin et al., 1989) but inhibits osteoblast proliferation at high concentrations. In osteoblast-rich cultures, PTH promotes proliferation (Van der Plas et al., 1985). These observations suggest that the anabolic effect of PTH is mediated by an increase in proliferation of osteoblast precursors. In most systems, the mitogenic effect of PTH is associated with a reduction in the expression of the osteoblast phenotype. It should be remembered, however, that the response of different osteoblast cultures to PTH is variable and depends on whether they are primary cultures or a cell line, on the species and age of the source of the osteoblastic cells, and on the initial degree of differentiation in vitro.

There is convincing evidence that PTH-stimulated bone formation is mediated by local growth factors such as IGF-I and -II and TGF-β. PTH stimulates production of IGF-I in isolated bone cells and in rodent calvarial organ cultures (Canalis et al., 1993). While PTH does not appear to regulate osteoblastic synthesis of TGF-β (McCarthy et al., 1989), its action to stimulate osteoclastic bone resorption results in liberation of TGF-β from bone matrix. The acidic microenvironment under the osteoclast may activate latent TGF-β, which would stimulate osteoblast chemotaxis, replication, and differentiation of osteoblasts. PTH also increases production of interleukin (IL)-6 in bone cells (Feyen et al., 1989), and plasma IL-6 concentrations are increased in patients with primary and secondary hyperparathyroidism (Rusinko et al., 1995).

b. *Calcitriol.* Calcitriol, or 1,25-dihydroxyvitamin D_3 (1,25-$(OH)_2D_3$), is produced in the proximal tubules of the kidney by the 1α-hydroxylation of 25-hydroxyvitamin D_3 and is a potent bone-resorbing factor. 1,25-$(OH)_2D_3$ increases intestinal absorption of calcium and phosphate, and its presence in vivo is necessary for normal bone mineralization. The production and effects of calcitriol are intimately related to those of PTH as PTH stimulates production in the kidney of 1α-hydroxylase, the rate-limiting enzyme needed for the conversion of 25-hydroxyvitamin D_3 to 1,25-$(OH)_2D_3$. Calcitriol receptors are located on osteoblasts and their precursors (Chen et al., 1983; Narbaitz et al., 1983). The effects of calcitriol on osteoblast-like cells in vitro are disparate and often do not correlate with what is known about its function *in vivo*. In short-term bone organ cultures, calcitriol inhibits bone collagen synthesis (Raisz *et al.*, 1980), and when incubated with cultured bone cells with the osteoblast phenotype, it increases expression of osteocalcin and alkaline phosphatase (Price and Baukol, 1980; Manolagas *et al.*, 1981). In prolonged primary cultures of fetal rat calvarial cells, 1,25-$(OH)_2D_3$ inhibited the formation of mineralized bone nodules, while in rodent osteoblasts and osteosarcoma lines, it inhibited osteoblast proliferation in a dose-dependent manner (Chen *et al.*, 1983). In contrast, calcitriol stimulates DNA synthesis in periosteal cells of calvaria (Canalis and Lian, 1985). Thus, PTH and calcitriol can influence proliferation of osteoblasts or their precursors, and the effects may depend on the stage of differentiation of the target cell population tested, as well as the cell system used.

c. Estrogen. Sex steroids play an important role in bone physiology through the maintenance of mineral homeostasis and bone balance. Sex steroid deficiency predisposes to bone loss and fracture. Not only is estrogen the most important sex steroid in preventing osteoporosis in women, but the finding of functional estrogen receptor deficiency in a young male patient with osteoporosis and failure to fuse epiphyseal growth plates suggests that estrogen is important in maintaining bone mass in men as well (Smith *et al.*, 1994).

Estrogen receptors have been demonstrated in cultured bone cells of the osteoblast lineage (Komm *et al.*, 1988; Eriksen *et al.*, 1988) as well as on osteoclasts (Oursler *et al.*, 1991b). The effects of estrogen on osteoblast-like cells, using a variety of *in vitro* models, have been conflicting (Turner *et al.*, 1994). Estrogen has been reported to increase (Ernst *et al.*, 1989), decrease (Gray *et al.*, 1987), or have no effect on (Keeting *et al.*, 1991) proliferation of cells of the osteoblast lineage from a variety of sources. Likewise, there are similar reports that estrogen increased (Gray *et al.*, 1987), decreased (Watts *et al.*, 1989), or had no effect on production of bone matrix proteins. It is likely that the differences in the model systems used to study the effects of estrogen on osteoblasts, and the fact that the cell culture systems do not mimic the environment *in vivo,* may explain some of these conflicting results. This is in contrast to results *in vivo,* when estrogen enhances bone formation.

Estrogens stimulate production of TGF-β in human osteoblast-like cells (Oursler *et al.*, 1991a). These effects may be mediated through the nuclear protooncogenes *c-fos* and *c-jun* (Turner *et al.*, 1994; Weisz and Rosales, 1990).

There is sufficient evidence to suggest that estrogen has important effects on bone by modulating the production of locally active growth factors and cytokines in bone. Estrogen may decrease bone resorption by mediating release of cytokines and soluble growth factors by osteoblasts. This is likely a complex process and may involve multiple cytokines and growth factors. In osteoblast-like cells, estrogen has been shown to increase TGF-β production; have no effect on IL-1β, IL-8, or granulocyte–macrophage colony-stimulating factor (GM-CSF) production; and have variable effects on production of IL-6, IGF-I, IGF-II, and tumor necrosis factor (TNF)-α (Turner et al., 1994).

d. Insulin. Insulin stimulates bone matrix synthesis and has a stimulatory effect on the differentiated function of the osteoblast. However, it does not increase the number of collagen-producing cells (Canalis, 1993). *In vivo,* insulin is necessary for the normal mineralization of bone. Patients with insulin deficiency and diabetes mellitus have a higher incidence of osteopenia, although the mechanisms responsible for this observation are not well understood.

e. Growth Hormone. Growth hormone (GH) is an important stimulator of linear growth and also regulates bone formation after the growing phase of the skeleton is complete. Patients with GH deficiency have low bone mass, while patients with acromegaly often have increased bone mass. The mechanisms by which GH exerts its effects on bone are unclear. It stimulates production of IGF-I by skeletal cells, but it likely has other effects independent of IGF-I, as the anabolic effects of PTH on bone are not observed in hypophysectomized rats until GH is administered (Hock and Fonseca, 1990).

f. Glucocorticoids. Glucocorticoids in excess profoundly decrease bone mass. Glucocorticoids inhibit gonadotropin and sex steroid production, and the resulting hypogonadism can result in osteopenia (Lukert and Raisz, 1990). In addition to these indirect effects, glucocorticoids inhibit bone formation and stimulate bone resorption *in vivo. In vitro,* glucocorticoids have complex effects on gene expression in osteoblasts. These actions are dependent on the stage of osteoblast growth and differentiation and on the cell model and culture conditions used. Glucocorticoids have been shown to induce cells of the osteoblast lineage to differentiate into mature cells expressing the osteoblast phenotype (Delany *et al.,* 1994a). Other effects of glucocorticoid on bone include enhanced PTH receptor expression, decreased replication of preosteoblasts, inhibition of type I collagen and osteocalcin expression, increased interstitial collagenase expression, and decreased tissue inhibitor of metalloproteinases expression.

Although glucocorticoids have no effect on TGF-β synthesis, they can shift the binding of TGF-β1 from signal-transducing receptors to non-signal-transducing receptors and decrease TGF-β mitogenic properties in bone cells (Centrella et al., 1992). Glucocorticoids also inhibit expression of IGF-1 but do not modify IGF-II synthesis by osteoblasts. These steroid hormones also modify IGF-binding protein (IGFBP) expression, another mechanism by which they can regulate IGF actions in bone. Six IGFBPs have been identified, and binding of these proteins to IGFs can enhance or inhibit the biological effects of IGFs. Osteoblasts express all six IGFBPs, and production varies with the cell system studied. Glucocorticoids decrease IGFBP-3, -4, and -5 in human osteoblasts. As IGFBP-5 is the only IGFBP known to enhance bone growth, this may be another possible mechanism by which glucocorticoids inhibit bone formation. The IGF and IGFBP effects are very complex, and further elucidation of the mechanisms by which glucocorticoids alter their interactions is an important area of future investigation.

2. Local Factors

a. Transforming Growth Factor-β. TGF-β is secreted by osteoblasts in a latent, biologically inactive form that is incorporated into bone ECM. Osteoblasts not only produce TGF-β but also possess high-affinity receptors for it (Robey et al., 1987), providing the opportunity for autocrine stimulation of osteoblast replication. Latent TGF-β can be activated *in vitro* by a number of agents, including acid pH, or by proteases such as plasmin or cathepsin (Lyons et al., 1988; Miyazono et al., 1988). TGF-β1 and -β2 are homologous disulfide-linked homodimers of 25 kDa that have powerful effects on bone. These growth-regulatory factors are present in the bone matrix in concentrations of 0.1 mg/kg dry weight.

TGF-β can regulate gene expression of other locally active growth factors and cytokines. The local function of TGF-β may be very important in contributing to the differentiated activity of osteoblasts. It stimulates collagen synthesis and regulates gene expression of mRNA for pro-α1(I)-chain collagen, osteonectin, alkaline phosphatase, fibronectin (Noda and Rodan, 1987), osteopontin (Noda et al., 1988), and osteocalcin (Noda, 1989). TGF-β increases the abundance of matrix proteins by stimulating their synthesis and inhibiting degradation. It is a potent stimulator of collagen and fibronectin synthesis and secretion in fibroblasts and osteoblasts (Ignotz and Massague, 1986; Sato et al., 1987), acting by increasing the mRNA for collagen and fibronectin (Ignotz et al., 1987). TGF-β also inhibits degradation of matrix proteins by decreasing the synthesis of matrix-degrading enzymes, as well as increasing the synthesis of protease inhibitors (Sporn et al., 1987).

TGF-β promotes the differentiation of cells of the osteoblast lineage toward the mature osteoblast and the formation of new bone.

Bone is the most abundant source of TGF-β in the body (Hauschka et al., 1986). TGF-β has been identified in culture medium conditioned by fetal rat calvaria (Centrella and Canalis, 1985), and confirmation of its osteoblast origin comes from the demonstration that fetal bovine osteoblasts transcribe TGF-β mRNA, synthesize and secrete the peptide, bear high-affinity cell surface receptors for it, and are mitogenically stimulated by TGF-β, probably through an autocrine mechanism (Robey et al., 1987). Stimulation of the replication of normal cells of the osteoblast lineage (Canalis, 1994) has been shown, as has growth inhibition of osteoblast-like osteogenic sarcoma cells (Noda and Rodan, 1987; Pfeilschifter et al., 1987) and of a clonal murine osteoblast-like cell line (Noda and Rodan, 1986). When fetal rat calvarial cells are exposed to active TGF-β, they respond with increased proliferation, while continued exposure impairs bone cell differentiation and the formation of mineralized nodules (Ghosh-Choudhury et al., 1994). The direction of TGF-β effects on proliferation of cells in the osteoblast lineage may depend on the stage of differentiation of the target cell as well as interactions with other factors.

In addition to cell replication, matrix formation, and bone-related protein synthesis, TGF-β has been shown to induce fetal rat osteoblasts to produce hematopoietic cytokines. TGF-β induces IL-6 secretion in fetal rat bone cell cultures and enhances production of GM-CSF stimulated by PTH, IL-1, and bacterial endotoxin lipopolysaccharide in such cultures (Centrella et al., 1994). Additionally, TGF-β has major effects on human osteoblast cell protooncogene (*c-fos* and *c-jun*) expression (Subramaniam et al., 1995).

TGF-β also increases the capacity of osteoblasts to migrate unidirectionally (Pfeilschifter et al., 1989b). These data suggest that it may have an important chemotactic function in normal bone remodeling, attracting osteoblast precursors to sites of active bone resorption.

TGF-β, injected subcutaneously adjacent to bone surfaces, causes a profound increase in new bone formation (Noda and Camilliere, 1989; Marcelli et al., 1990; Mackie and Trechsel, 1990). When TGF-β is administered by injection over the calvaria of mice daily for 3 days, bone width is increased 40% over the next month (Marcelli et al., 1990). This is initially woven bone, but it is later replaced by lamellar bone. Similar effects are seen when TGF-β is injected or infused directly into the marrow cavity of the femur.

b. Insulin-like Growth Factors I and II. IGF-I and -II are weak bone cell mitogens but have clear and potent stimulatory effects on the differentiated function of the osteoblast, as evidenced by an increase in osteocalcin and type I collagen synthesis in osteoblasts. As a result, IGFs increase bone matrix apposition rates and bone formation. IGFs also decrease collagen degradation and the expression of interstitial collagenase, suggesting a role in the preservation of bone matrix. IGFs enhance bone formation *in vivo* (Spencer et al., 1993). Mice with null mutation of type I IGF receptor have delayed skeletal development and delayed ossification (Liu et al., 1993). The anabolic properties if IGF-I and -II, their inhibitory actions on matrix degradation, and their abundance in bone tissue suggest that these factors play a central role in the maintenance of bone mass (Canalis, 1994).

Several systemic factors can regulate IGF production. The anabolic effect on bone of intermittent administration of PTH has been attributed to expression of IGF-I. In organ cultures, transient exposure to PTH enhanced the production of IGF-I by three- to fourfold. Skeletal IGF production is only modestly dependent on growth hormone (Canalis, 1994), as growth hormone only marginally stimulates IGF-I production in osteoblasts (Delany et al., 1994b). 17β-Estradiol can increase IGF-I transcription in cells of the osteoblast lineage, while calcitriol has variable effects that differ with the model system studied. Glucocorticoids decrease IGF-I transcripts and protein concentrations in primary osteoblast cultures. Prostaglandin E_2 (PGE_2) has been shown to increase IGF-I synthesis in osteoblast cultures, and promoter analysis of the IGF-I promoter revealed PGE_2-responsive elements (Pash et al., 1995). In contrast, TGF-β, PDGF, and FGF inhibit IGF-I and -II expression. Neither TGF-β, PDGF, or FGF stimulate differentiation of osteoblasts, while BMPs are potent stimulators of osteoblast differentiation. It is interesting to note that BMPs increase IGF-I production, suggesting that the ability of BMPs to stimulate IGF-I production (Canalis and Gabbitas, 1994) is linked to their ability to induce differentiation in osteoblasts.

Regulation of IGF-I in bone is further complicated by the production of IGFBPs by osteoblasts. Osteoblasts express transcripts for the six known IGFBPs. Binding of IGF to one of these binding proteins can inhibit or potentiate the biological effect of IGF. Binding to IGFBP-1, for example, decreases the biological activity of IGF-I. Conversely, IGFBP-5 has been shown to increase bone formation and, thus, appears to enhance the effect of IGF-I. These observations are further complicated by the observation that growth factors such as TGF-β, PDGF, FGF, and BMP-2 inhibit synthesis of IGFBP-5 in bone cell cultures (Canalis and Gabbitas, 1995; Gabbitas and Canalis, 1995), while IGF-I and retinoic acid increase it (Dong and Canalis, 1995).

Thus it appears that the effect of IGFs on bone is anabolic and that local regulation of IGFs in bone is highly complex.

c. *Bone Morphogenetic Proteins.* BMPs are bone-derived peptides and members of the extended TGF-β superfamily. At least eight members are currently recognized. BMP-2 through BMP-8 share some TGF-β-related gene sequences. BMPs are synthesized by bone cells locally and stimulate the formation of ectopic bone when injected intramuscularly or subcutaneously into rodents (Urist, 1965). BMPs stimulate the replication and differentiation of normal cells of the osteoblast lineage and, in contrast to TGF-β, enhance the expression of the differentiated osteoblastic phenotype (Harris *et al.*, 1995; Canalis, 1994). BMP-1, -2, -3, -4, and -6 are temporally expressed in primary cultures of fetal rat calvarial cells (Ghosh-Choudhury *et al.*, 1994). BMP-2, -4, and -7 have been shown to induce differentiation of primitive mesenchymal cells into bone when implanted into subcutaneous tissue (Harris *et al.*, 1995). BMP-2 accelerates differentiation in primary cultures of fetal rat calvarial cells, as demonstrated by an increase in expression of alkaline phosphatase and osteocalcin (Harris *et al.*, 1995). BMP-3 decreases osteoclastic bone resorption and is chemotactic for monocytes (Cunningham *et al.*, 1992). BMP-7 (osteogenic protein-1) suppresses cell proliferation and stimulates the expression of markers characteristic of the osteoblast phenotype in rat osteosarcoma cells, but stimulates growth and differentiation in rat calvarial cultures (Maliakal *et al.*, 1994). The precise role of BMPs in the bone remodeling process has yet to be determined.

d. *Heparin-Binding Fibroblast Growth Factors.* Both acidic and basic FGFs are present in mineralized bone matrix and stimulate the replication of cells in the skeletal system, but do not increase the differentiated function of the osteoblast. Therefore, they may play an important role in bone repair when bone cell mitogenesis may be necessary (Canalis, 1994). TGF-β has been demonstrated to increase basic FGF expression in a mouse osteoblast cell line, while PTH and IL-1 had no effect (Hurley *et al.*, 1994). Acidic and basic FGF have powerful stimulatory effects on bone formation *in vivo*. When injected locally over the calvaria of mice, FGF causes a 50% increase in bone thickness. When administered to ovariectomized rats, FGF blocked the associated bone loss and also increased trabecular connectivity and bone microarchitecture (Dunstan *et al.*, 1995).

e. *Platelet-Derived Growth Factor.* PDGF stimulates the replication of bone cells but does not increase the differentiated function of the osteoblast (Hock and Canalis, 1994; Canalis, 1994). PDGF is produced by cells with the osteoblast phenotype (Graves and Owen, 1983; Graves *et al.*, 1984), and osteoblasts have PDGF receptors (Xie *et al.*, 1994). The A-chain homodimer of PDGF (PDGF-AA) is a less potent mitogen for osteoblasts cultured from fetal rat bone than those isoforms containing the B-chain subunits (PDGF-BB). Osteoblasts only synthesize PDGF A subunits, while the systemic form contains only PDGF B chains (Canalis *et al.*, 1992). While some investigators have found that IL-1α and TNF-α synergistically enhance the mitogenic effect of PDGF-AA, but not PDGF-BB, on osteoblasts, partly through enhanced binding (Centrella *et al.*, 1992), others have demonstrated that IL-1 reduces PDGF receptor expression (Yeh *et al.*, 1993) and reduces PDGF-AA binding to osteoblasts (Gilardetti *et al.*,1991). Since PDGF is a major component of platelets and is secreted during the platelet release reaction, its production at sites of fracture healing may be important to the regeneration of bone.

f. Prostaglandins. Prostaglandins have multiple effects on cells in the osteoblast lineage. They inhibit calvarial collagen synthesis in short-term cultures (Dietrich *et al.*, 1976a,b), and, with long culture periods, collagen synthesis increases (Chyun and Raisz, 1984). Dogs injected locally with prostaglandins have increased bone volume (Ueno *et al.*, 1985). PGE_2 stimulates the replication of cells in the periosteum and promotes collagen synthesis (Chyun *et al.*, 1984), and increased collagen synthesis in response to PGE_2 has also been shown in clonal osteoblastic cells (Hakeda *et al.*, 1985). *In vivo* data support a stimulatory effect of prostaglandins on bone formation. The production of prostaglandins, predominantly PGE_2, by osteoblasts has been well documented (Noda *et al.*, 1982). PGE_2 increases IGF-I synthesis by osteoblasts as well (Pash *et al.*, 1995).

g. Interleukin-6. Murine stromal cells, human bone cells, and rat and mouse osteoblast-like cell lines activated with IL-1 and TNF-α all secrete IL-6 (Horowitz, 1993; Chaudhary *et al.*, 1992), and IL-6 production has been demonstrated in normal human osteoblasts (Birch *et al.*, 1993; Chaudhary *et al.*, 1992). The secretion of IL-6 induced by these cytokines can be inhibited by treating the cells *in vitro* with estrogen and, to a lesser extent, with testosterone and progesterone (Girasole *et al.*, 1992). Other investigators, however, have seen variable effects of sex steroids on IL-6 production by osteoblasts and stromal cells in similar experiments (Chaudhary *et al.*, 1992; Rickard *et al.*, 1992). Thus, it is possible that cytokine secretion by bone cells is downregulated by the addition of estrogen directly to osteoblasts and stromal cells, a finding that is consistent with the expression of estrogen receptors by osteoblasts (Komm *et al.*, 1988; Eriksen *et al.*, 1988).

IL-6 production is also increased in bone cells by PTH (Feyen *et al.*,

1989), IL-1, or TNF (Littlewood *et al.*, 1991), suggesting that it may be a major mediator of osteoblast function. Specifically, PTH has been shown to induce IL-6 mRNA expression and protein secretion from a mouse osteoblast cell line as well as from primary rat osteoblast cultures (Greenfield *et al.*, 1993). IL-6 can stimulate human osteoclastic bone resorption, and possibly the stimulatory effects on PTH, 1,25-$(OH)_2D_3$, and IL-1 on bone resorption may in part be mediated through IL-6 production by osteoblasts.

h. Tumor Necrosis Factor-α. TNF-α is a cytokine produced by macrophages and monocytes that has been implicated in the pathogenesis of osteoporosis, cancer-related cachexia, and septic shock. TNF-α stimulates bone resorption *in vitro* and causes hypercalcemia when administered to mice. The effects of TNF-α on bone formation are not as well understood and, from available literature, appear to be inhibitory. *In vitro,* TNF-α treatment of human osteoblasts inhibits expression of alkaline phosphatase and decreases collagen incorporation into developing matrix as well as mineralization of matrix without an effect on DNA synthesis (Panagakos *et al.*, 1994). TNF-α treatment of various osteoblast-like cell lines inhibits PTH-stimulated rise in cAMP and free cytosolic calcium (Hanevold *et al.*, 1993; Katz *et al.*, 1992) by downregulating PTH receptors (Schneider *et al.*, 1991). TNF-α inhibits calcitriol-stimulated synthesis of osteocalcin and inhibits vitamin D receptor number in a clonal rat osteosarcoma cell line (Mayur *et al.*, 1993). *In vivo,* infusion of TNF-α inhibits new bone formation stimulated by PTH-related protein (PTHrP) (Barengolts *et al.*, 1994). Taken together, these observations indicate that TNF-α has an inhibitory effect on bone formation and may be one of the factors responsible for the uncoupling of bone resorption and bone formation observed in malignancy.

i. Interleukin-1. IL-1 has growth-stimulating effects on osteoblast cells but inhibits differentiated function (Gowen and Mundy, 1986). IL-1 along with TNF impaired the responsiveness to PTH and PTHrP in cultures of clonal rat osteosarcoma cells (UMR-106) via downregulation of PTH receptors (Katz *et al.*, 1992). Additionally, IL-1 and TNF-α had variable effects on osteoblastic expression of osteocalcin, alkaline phosphatase, and mineralized matrix depending on the system studied (Taichman and Hauschka, 1992).

j. Interleukin-1 Receptor Antagonist. The IL-1 receptor antagonist (IL-1ra) is a naturally occurring antagonist to IL-1α and -1β that mediates its effects via binding to the IL-1 receptor (Hannum *et al.*, 1990; Eisenberg *et al.*, 1990). IL-1ra decreases bone loss in ovariectomized rats (Kimble *et al.*, 1994), and clinical studies suggest that secretion of

IL-1ra in the postmenopausal state may limit bone loss by decreasing IL-1 bioactivity (Pacifici et al., 1993).

k. *Interleukin-4.* IL-4, a cytokine produced mainly by T lymphocytes, modulates the activity of lymphoid, hematopoietic, and mesenchymal cells. In vitro, IL-4 inhibits proliferation of osteoblast-like cells as well as enhances the expression of alkaline phosphatase stimulated by calcitriol (Riancho et al., 1993b). IL-4 alone inhibits expression of alkaline phosphatase and slows matrix mineralization in MC3T3 cells and primary cultures of rodent osteoblasts (Riancho et al., 1995). IL-4 stimulates monocyte colony-stimulating factor (M-CSF) expression, but not IL-1, IL-6, GM-CSF, or PGE_2, by MC3T3 cells in these studies. In contrast, IL-4 increased hydroxyproline and osteocalcin accumulation and caused mineralization in cultured human osteoblast-like cells (Ishibashi et al., 1995). Additionally, IL-4 mediated osteoblast migration in a modified Boyden chamber system (Lind et al., 1995). In vivo, transgenic mice that inappropriately express IL-4 under the direction of the lymphocyte-specific proximal promoter for the *lck* gene display severe osteoporosis of both cortical and trabecular bone (Lewis et al., 1993). The observed osteoporosis is characterized by decreased bone formation. Despite the observed disparate effects of IL-4 on osteoblast function in various *in vitro* systems, the *in vivo* results in transgenic mice suggest that the major effect on bone is to inhibit osteoblast function.

l. *Interleukin-11.* IL-11 is a pleomorphic cytokine with biological activities that overlap with those of IL-6. Human osteosarcoma (SaOS-2) cells and primary human osteoblasts produce IL-11. Its production is enhanced in SaOS-2 cells after stimulation with IL-1, TGF-β, PTH, and PTHrP but not with IL-4, interferon-γ or endotoxin (Elias et al.,1995). IL-11, along with IL-6, may be an important component of the cytokine network mediating osteoblast–osteoclast communication in bone remodeling (Manolagas and Jilka, 1995).

m. *Parathyroid Hormone-Related Protein.* Tumor-produced PTHrP has an established role as a mediator of malignancy-associated hypercalcemia. Its role in normal physiology is still under intense investigation, and studies demonstrate that PTHrP is important in normal bone development. Mice homozygous for the null mutation for the PTHrP gene are born with premature mineralization of normally cartilaginous areas (Karaplis et al., 1994). Furthermore, transgenic mice with targeted overexpression of PTHrP to proliferating and prehypertrophic chondrocytes by the mouse collagen type II promoter are born with marked foreshortening of the limbs and tail. Histological analysis of affected bones revealed a marked delay in chondrocyte maturation and endochondral ossification (Weir *et al.*, 1995). In other experiments, PTHrP was detected in conditioned media from cultures of normal human bone cells, and PTHrP mRNA was detected in cells of the osteoblast lineage (Walsh *et al.*, 1995). PTHrP production was inhibited by a variety of glucocorticoids in this system (Walsh *et al.*, 1995). Taken together, these findings indicate that PTHrP is produced by normal bone cells and likely has an important role in the development of the normal skeleton. Determining this role is an important area of future investigation.

III. Osteoclasts

A. General Characteristics of the Osteoclast

Osteoclasts are large, multinucleated cells that are the primary bone-resorbing cells. They are hematopoietic in origin and are formed by fusion of mononuclear precursors in the marrow. The osteoclast contains large quantities of tartrate-resistant acid phosphatase, a marker enzyme for the osteoclast, as well as hydrolytic enzymes, which are involved in the bone-resorptive process. Osteoclast formation and activity are regulated by factors produced from cells in the bone marrow microenvironment, including osteoblasts, stromal cells, and monocyte–macrophages. Characteristic features of the osteoclast include the presence of a ruffled border adjacent to areas in which bone is resorbed, as well as pleomorphic mitochondria and a prominent Golgi apparatus. In addition, osteoclasts have several unique features that distinguish them from macrophage polykaryons. These include the presence of calcitonin receptors, their ability to contract in response to calcitonin, cross-reactivity with several osteoclast-specific antibodies such as 121F (described by Oursler *et al.*, 1985), and absence of Fc receptors.

As noted earlier, osteoclasts are formed from hematopoietic precursors. This is based on transplantation studies in osteopetrotic rodents as well as in humans (Walker, 1972, 1973; Coccia *et al.*, 1980; Sorell *et al.*, 1981), and in studies using bone marrow culture techniques (MacDonald *et al.*, 1987; Takahashi *et al.*, 1988; Kurihara *et al.*, 1990b). The leading candidate for the earliest identifiable osteoclast precursor is the colony-forming unit–granulocyte–macrophage (CFU-GM), the granulocyte–macrophage progenitor. This cell, under the influence of a variety of cytokines, appears to differentiate to a more committed unipotent osteoclast precursor, which expresses calcitonin receptors

(Takahashi et al., 1995a). These cells then fuse to form the multinucleated osteoclasts. The committed osteoclast precursor is postmitotic, expresses tartrate-resistant acid phosphatase, and cross-reacts with monoclonal antibodies that identify osteoclasts.

Osteoclasts resorb bone through the production of proteolytic enzymes and secretion of hydrogen ions into the localized microenvironment under the ruffled border. This extracellular lysosome that is formed beneath the ruffled border results in degradation of collagen and calcified matrix. Hydrogen ion production in the osteoclast appears to be generated by the enzyme carbonic anhydrase II, and these hydrogen ions are then pumped across the ruffled border by a vacuolar-type proton pump (Blair et al., 1989). Lysosomal enzymes are also released by the osteoclast and are highly activated in this acidic microenvironment. The bone resorption process results in formation of a resorption lacuna underneath the osteoclast. As the osteoclast then moves across the bone surface, additional resorption lacunae can be formed by a single osteoclast. Any process that interferes with the molecular mechanisms responsible for bone resorption or osteoclast formation results in osteopetrosis.

A variety of techniques have been used to study osteoclast formation and function, including isolation of authentic osteoclasts from long bones, use of marrow culture techniques in which cells with an osteoclast phenotype form, and use of giant cells isolated from giant cell tumors of bone as models for highly activated human osteoclast-like cells (Mundy and Roodman, 1987). Methods such as these have been employed to characterize osteoclasts and their precursors, as well as to determine the effects of local factors and systemic hormones on osteoclast formation and osteoclast activity. In addition, bone organ culture systems have been very useful for identifying factors controlling osteoclastic bone resorption, and several laboratories, including those of Lowik et al. (1989) and Burger and her associates (1982), have used fetal bone rudiments to study osteoclast biology and recruitment. Development of transgenic mice and mice in which gene function has been ablated by the techniques of homologous recombination has further helped to identify the role of specific factors in osteoclast activity and osteoclast formation. Studies in animals with osteopetrosis have also demonstrated some of the molecular defects responsible for impaired osteoclast function. These model systems have shown the importance of both locally acting factors and systemic hormones on osteoclast function and helped to elucidate the role that these factors play in both normal and pathological states, such as osteoporosis, Paget's disease of bone, and multiple myeloma.

B. Systemic Hormones That Affect Osteoclast Function and Formation

1. *Parathyroid Hormone*

As noted earlier, PTH can either induce bone formation or inhibit osteoblast activity. When it is administered continuously, bone formation is suppressed and osteoclastic bone resorption is increased. Osteoclast precursors and osteoclasts have been shown to express PTH receptors (Teti *et al.*, 1991; Hakeda *et al.*, 1989), but the major mechanism that has been proposed for the action of PTH on osteoclasts has been an indirect one, in which PTH acts on the osteoblast, which in turn stimulates osteoclastic bone resorption. In patients with hyperparathyroidism, there is a loss of bone due to increased osteoclastic bone resorption accompanied by marrow fibrosis. PTH causes a marked increase in bone resorption in bone organ culture systems and stimulates osteoclast formation in both murine and human marrow culture systems (Takahashi *et al.*, 1988; MacDonald *et al.*, 1987). In addition, Lorenzo *et al.* (1986) have shown that PTH acts predominantly on a postmitotic cell. We have described an *in vivo* model of osteoclast formation and demonstrated that PTH and PTHrP appear to act on the more differentiated osteoclast precursor, rather than the proliferative early osteoclast precursor (Uy *et al.*, 1995b). This more differentiated precursor then fuses to form osteoclasts. Cytokines such as IL-1, TGF-α, TNF-α, and IL-6 can enhance the effects of PTHrP on osteoclast formation and osteoclastic bone resorption. IL-6, for example, appears to stimulate proliferation of early osteoclast precursors, and PTHrP then induces the differentiation and fusion of these precursors to form multinucleated osteoclasts (De La Mata *et al.*, 1995). Furthermore, PTH and PTHrP enhance calcium reabsorption by the kidney. This enhanced renal calcium reabsorption contributes to the hypercalcemia seen in patients with malignancies, who have increased osteoclastic bone resorption due to overproduction of PTHrP by their tumors. Thus, although it is unclear if PTH acts directly or indirectly on osteoclasts, it has major effects on osteoclast formation and osteoclastic bone resorption in both normal and pathological states.

2. *Calcitriol*

Vitamin D_3 and its metabolites are potent stimulators of osteoclastic bone resorption and osteoclast formation. The most active metabolite, $1,25\text{-}(OH)_2D_3$, acts as a fusigen for committed osteoclast precursors (Takahashi *et al.*, 1987). $1,25\text{-}(OH)_2D_3$ does not act on mature osteoclasts directly, since the mature osteoclast appears to lack vitamin D receptors (Narbaitz *et al.*, 1983), and thus may have indirect effects on osteoclastic bone resorption (Feyen et al., 1989). 1,25-$(OH)_2D_3$ can induce IL-1 and IL-6 production by osteoblasts, factors that stimulate osteoclastic bone resorption. Furthermore, 1,25-$(OH)_2D_3$ increases calcium absorption from the gut and can act in conjunction with PTH to stimulate renal tubular calcium reabsorption, as well as enhance osteoclastic bone resorption stimulated by PTH.

3. Prostaglandins

Prostaglandins are potent stimulators of osteoclastic bone resorption in bone organ culture systems and stimulate osteoclast formation in murine marrow cultures (Takahashi et al., 1988). However, PGE_2 inhibits osteoclastic bone resorption and formation in human systems (Chenu et al., 1990; Chambers et al., 1985). Chambers et al. (1985) have reported that PGE_2 induces osteoclastic contraction analogous to calcitonin and inhibits bone resorption by isolated osteoclasts. The effects of prostaglandins on osteoclast formation and osteoclastic bone resorption may be dose dependent and dependent on the assay system used. Tashjian and associates (1985) have found that a variety of factors that stimulate osteoclastic bone resorption in the mouse calvarial organ culture system do so by generating prostaglandins. Recently, Gallwitz et al. (1993) have shown that other arachidonic acid metabolites, such as the peptidoleukotrienes E_1 and D_4, as well as 5-hydroxyeicosatetraenoic acid, stimulate isolated osteoclasts to resorb bone. These arachidonic acid metabolites may play an important role in bone resorption in areas of chronic inflammation.

4. Calcitonin

Calcitonin is a peptide hormone secreted by the parafollicular cells of the thyroid gland and is a potent inhibitor of osteoclastic bone resorption. It acts at multiple stages in the osteoclast lineage, including inhibition of osteoclast formation as well as the bone-resorbing capacity of mature osteoclasts. Calcitonin receptors are expressed on committed osteoclast precursors and appear to be a differentiation marker for the mature osteoclast (Takahashi et al., 1995a). Calcitonin downregulates expression of calcitonin receptors and osteoclast precursors in mature osteoclasts by inhibiting expression of the messenger RNA for its receptor (Takahashi et al., 1995a). Calcitonin acts on osteoclasts by stimulating adenylcyclase activity and cAMP accumulation, which results in immobilization of the osteoclasts and their contraction away from the bone surface. Osteoclasts continually exposed to calcitonin escape from its effects. The mechanism responsible for this is unclear, but it may be due to the effects of calcitonin on expression of the calcitonin receptor. Calcitonin has been used as a therapeutic agent in patients with Paget's disease and the hypercalcemia of malignancy, as well as those with osteoporosis. In contrast to patients with hypercalcemia of malignancy, patients with Paget's disease appear to be capable of responding to calcitonin for prolonged periods of time. However, the molecular mechanisms responsible for the increased responsivitiy of pagetic osteoclasts to calcitonin have yet to be defined.

C. Autocrine–Paracrine Factors with Osteoclast-Stimulatory Activity

1. *Interleukin-1*

IL-1 is a cytokine produced by monocyte–macrophages and marrow stromal cells. It can stimulate bone resorption *in vitro* and *in vivo*. Several authors have shown that IL-1 induces bone resorption and osteoclast-like cell formation in murine and human marrow cultures (Gowen *et al.*, 1983; Pfeilschifter *et al.*, 1989a).

Uy *et al.* (1995a) have used an *in vivo* model of osteoclast formation to examine the effects of IL-1 on the different stages of osteoclast differentiation. They used IL-1 as an osteotropic factor to delineate the cellular mechanisms responsible for enhanced osteoclast activity stimulated by IL-1. IL-1 induced hypercalcemia and enhanced the growth and differentiation of CFU-GM, the earliest identifiable osteoclast precursor; increased the number of more committed mononuclear osteoclast precursors; and stimulated mature osteoclasts to resorb bone.

The effects of marrow stromal cells on osteoclast formation are also in part mediated by IL-1. Takahashi *et al.* (1995b) established a human bone marrow stromal cell line (Saka cells) by infecting marrow-adherent cells from semisolid marrow cultures with a recombinant adeno/simian virus-40 virus. Coculture of Saka cells with human marrow mononuclear cells enhanced formation of osteoclast-like multinucleated cells in human marrow cultures. These osteoclasts expressed calcitonin receptors and formed resorption lacunae on dentine. Polymerase chain reaction analysis of the Saka cells detected expression of mRNAs for IL-1β. Addition of neutralizing antibodies to IL-1β blocked the effects of Saka cells on osteoclast-like cell formation.

IL-1 has been implicated in the increased bone loss seen in several pathological states. It is produced by several tumors associated with hypercalcemia, such as squamous cell carcinoma and lymphoma (Fried *et al.*, 1989; Sato *et al.*, 1989). Freshly isolated marrow cells derived from some patients with myeloma produced IL-1β, and the bone-resorbing activity present in culture media from these marrow cell isolates could be neutralized by IL-1β antibodies (Kawano et al., 1989; Cozzolino et al., 1989).

Kitazawa et al. (1994) have demonstrated that IL-1 plays a direct role in mediating the effects of ovariectomy on osteoclastogenesis and bone resorption. Ovariectomy increased the bone marrow cell secretion of IL-1 and the formation of TRAP(+) osteoclast-like cells in bone marrow cultures treated with vitamin D_3 and in vivo. The effects of ovariectomy on osteoclast formation in vivo and in vitro were blocked by treatment with an anti-IL-1ra.

The mechanisms mediating the effects of IL-1 on osteoclasts are not completely understood. In an in vivo model system designed by Boyce et al. (1989a,b), administration of IL-1α and IL-1β systemically stimulated bone resorption in mice. Treatment of the mice with indomethacin partially inhibited the effects of IL-1, suggesting that part of the effects of IL-1 were mediated by prostaglandin. Thus, IL-1 is a potent osteotropic factor that stimulates osteoclasts at all stages of differentiation and induces bone resorption both in vivo and in vitro. Its role in disease states, as well as the biology of its effects on the osteoclasts, remain to be completely elucidated.

2. Colony-Stimulating Factors

Colony-stimulating factors are hematopoietic growth factors that induce clonal growth of hematopoietic progenitors in vitro and in vivo. They are produced by macrophages, stromal cells, endothelial cells, and T lymphocytes in the marrow microenvironment. Since the osteoclast is hematopoietic in origin, it is not surprising that these factors may act as stimulatory factors for the osteoclast as well. Studies in animals and patients with osteopetrosis have shown that bone marrow transplantation can cure osteopetrosis (Walker, 1975a,b; Coccia et al., 1980; Sorell et al., 1981).

The op/op mouse, in which the M-CSF gene is mutated, has greatly reduced numbers of macrophages and osteoclasts. In vivo, exogenous M-CSF has been shown to be necessary for the generation of osteoclast-like cells in cocultures of hematopoietic and stromal cells from op/op osteopetrotic mice. In vivo administration of M-CSF restores osteoclastogenesis and bone resorption in op/op mice (Morohashi et al., 1994).

The effects of other colony-stimulating factors, including GM-CSF, are not clear as those for M-CSF. GM-CSF stimulates the growth of osteoclast precursors and induces osteoclast formation when human bone marrow cultures are treated sequentially with GM-CSF followed by $1,25(OH)_2D_3$ (MacDonald et al., 1986). However, GM-CSF by itself has no effect on osteoclastic bone resorption, and it has been reported to inhibit osteoclastic bone resorption in murine marrow cultures (Shuto et al., 1994). Takahashi and associates (1991) have examined the effects of colony-stimulating factors on murine osteoclast formation *in vitro* in a coculture system, and have shown that all these factors can stimulate the growth of osteoclast precursors but, when added simultaneously with $1,25(OH)_2D_3$, inhibit its effects on osteoclast formation.

3. *Transforming Growth Factor-α*

TGF-α is a polypeptide of 5700 Da that is partially homologous to epidermal growth factor (EGF). It is produced by several solid tumors associated with the hypercalcemia of malignancy and can stimulate osteoclastic bone resorption in murine organ cultures (Ibbotson et al., 1985). The effects of TGF-α are mediated through the EGF receptor and are independent of prostaglandin synthesis (Yates et al., 1992).

TGF-α induces osteoclastic bone resorption and hypercalcemia in nude mice and increases osteoclast-like cell formation in human marrow cultures (Yates et al., 1992; Takahashi et al., 1986a). Hiraga et al. (1995) demonstrated the presence of TGF-α by immunohistochemistry in areas of bone adjacent to osteoclast activation and bone resorption induced by a metastatic human melanoma cell line in nude mice. Therefore, TGF-α can stimulate osteoclast formation and bone resorption both *in vitro* and *in vivo*.

4. *Tumor Necrosis Factor*

Both TNF-α and TNF-β (lymphotoxin) markedly stimulate the formation of osteoclast-like multinucleated cells in human marrow cultures (Pfeilschifter et al., 1989a). TNF can also affect the activity of mature osteoclasts. Thomson et al. (1986, 1987) have shown stimulation of bone resorption by mature osteoclasts when these cells were incubated with IL-1 or TNF and cocultured with osteoblastic cells.

TNF potentiates the effects of IL-1 on osteoclast formation (Pfeilschifter et al., 1989a); therefore, its effects on osteoclasts may be in part mediated by other cytokines. Garrett et al. (1987) demonstrated that myeloma cell lines produced lymphotoxin *in vitro*, and neutralizing antibodies to lymphotoxin blocked the bone resorption in bone organ cultures induced by media conditioned by these cells. However, increased levels of TNF-β have not been found in an *in vivo* model of human myeloma bone disease (Alsina et al., 1995). TNFs appear to stimulate both proliferation and differentiation of precursors for osteoclast-like cells to osteoclasts and may be involved in the pathogenesis of hypercalcemia of malignancy.

5. Interleukin-6

IL-6 is a 26,000-Da cytokine produced by marrow stromal cells, monocyte–macrophages, osteoclasts, and osteoblasts. It induces osteoclast formation and bone-resorbing activity of preformed osteoclasts (Lowik et al., 1989; Kurihara et al., 1990a). However, it does not appear, by itself, to be a potent osteotropic factor in murine systems in vivo. IL-6 potentiates the effect of other hormones such as PTHrP on calcium homeostasis and osteoclastic bone resorption in vivo, as demonstrated by De La Mata et al. (1995).

IL-6 production by bone and marrow stromal cells is suppressed by 17β-estradiol in vitro. In mice, estrogen loss with ovariectomy increased the number of CFU-GMs, enhanced osteoclast development in ex vivo cultures of marrow, and increased the number of osteoclasts in trabecular bone. These changes were prevented by administration of an antibody to IL-6 (Jilka et al., 1992). These findings suggest that IL-6 may be involved in the increased bone resorption in postmenopausal osteoporosis, with estrogen loss resulting in an IL-6-mediated stimulation of osteoclasts. However, others have implicated IL-1 and/or TNF-α as potential mediators for the bone loss seen with ovariectomy (Kimble et al., 1994).

IL-6 may also act as an autocrine–paracrine factor in Paget's disease (Roodman et al., 1992). Osteoclast-like multinucleated cells formed in human marrow cultures from patients with Paget's disease actively release IL-6 into their conditioned medium, and this medium stimulated osteoclast-like cell formation in normal marrow cultures. Furthermore, patients with Paget's disease, but not normal subjects, have elevated levels of IL-6 in their marrow plasma and their peripheral blood.

We examined the effects of antisense constructs to IL-6 on the bone-resorbing capacity of purified giant cells from giant cell tumors of bone to help define the role of IL-6 in human osteoclastic bone resorption (Reddy et al., 1994). IL-6 levels were elevated in conditioned medium from highly purified giant cells. Treatment of these giant cells with IL-6 antisense constructs or neutralizing antibodies to IL-6 caused a fourfold decrease in IL-6 levels, and significantly decreased the number of resorptive lacunae formed and the area of the dentine resorbed. These observations demonstrate that IL-6 may play an important role in the bone-resorptive process of human osteoclasts.

Interestingly, IL-6 has been found to induce a marked loss of calcitonin-binding sites on normal T lymphocytes at concentrations known to be active on bone metabolism (Body et al., 1994). IL-6 may also play an important role in other states of increased bone destruction, such as multiple myeloma bone disease and Gorham-Stout disease, or disappearing bone syndrome, where it has been implicated by Devlin et al. (1995a).

6. *Annexin II*

Identification of the production of autocrine factors by osteoclasts represents an important addition to our understanding of normal osteoclast formation and activity. Takahashi et al. (1994) have prepared a mammalian cDNA expression library generated from highly purified human osteoclast-like multinucleated cells formed in long-term bone marrow cultures and screened this library for autocrine factors that enhance osteoclast-like cell formation. In the initial screening annexin II was identified, and purified recombinant annexin II significantly increased osteoclast-like cell formation in human bone marrow cultures in the absence of $1,25(OH)_2D_3$. It also enhanced the bone-resorptive capacity of $1,25(OH)_2D_3$ in bone organ cultures. Interestingly, annexin II mRNA was expressed at high levels in RNA isolated from highly purified giant cells from osteoclastomas, human osteoclast-like cells, and pagetic bone.

Nesbitt and Horton (1995) have reported that annexin II is also expressed on the surface of osteoclasts, and inhibition of annexin II with an antibody to it blocked bone resorption by isolated osteoclasts. Devlin and associates (1995b) have demonstrated that annexin II stimulates the proliferation of osteoclast precursors in human marrow cultures and enhances the effects of GM-CSF on the growth of these precursors. Annexin II appears to be a proliferative factor that enhances the growth of osteoclast precursors, as well as plays a role in osteoclastic bone resorption.

D. LOCAL INHIBITORY FACTORS

1. *Transforming Growth Factor-β*

TGF-β is one of the key factors involved in coupling bone formation to previous bone resorption (Mundy, 1991). It is secreted by osteoblasts and osteoclasts and may act as an autocrine factor stimulating osteoblastic bone formation through enhanced chemotaxis, proliferation, and differentiation of committed osteoblasts.

TGF-β is secreted as a dimer composed of 12.5-kDa subunits noncovalently associated with one or more polypeptides to form a higher molecular weight latent complex. Latent TGF-β can be experimentally activated by proteinase treatment or denaturation to remove the binding proteins.

Oursler (1994) has shown that osteoclasts expressed mRNA for TGF-β and that latent TGF-β that is secreted may be activated by osteoclasts. She concluded that osteoclasts may secrete proteinases that activate latent TGF-β into the extracellular space, and that TGF-β may be an autocrine factor for osteoclasts as well. Similarly, Pfeilschifter and Mundy (1987) have reported that osteoclasts can activate latent TGF-β.

TGF-β has been shown to be a potent inhibitor of osteoclastic bone resorption by modulating both osteoclast migration and osteoclast differentiation (Chenu et al., 1988). It probably also plays a role in the regulation of the proliferation of osteoclast progenitors *in vivo*, since it was shown by Dieudonne et al. (1991) to prevent the increase in the number of TRAP(+) multinucleated osteoclast-like cells when administered systemically to ovariectomized rats. Similarly, Chenu et al. (1988) have shown that TGF-β inhibits both the proliferation and fusion of human osteoclast precursors.

2. γ-Interferon

γ-Interferon is a potent inhibitor of bone resorption *in vitro* (Gowen and Mundy, 1986), and suppresses the formation and maturation of osteoclasts (Takahashi et al., 1986b). Tohkin et al. (1994) examined the effects of γ-interferon on humoral hypercalcemia in nude mice bearing lower jaw tumors, in which PTHrP is responsible for inducing hypercalcemia. Mice were injected with γ-interferon for 5 days before the establishment of hypercalcemia, and the increase in plasma calcium concentration was delayed. γ-Interferon also abolished the formation of multinucleated osteoclast-like cells from bone marrow cells of these mice *in vitro*. The data suggest that γ-interferon suppresses the formation of osteoclasts, resulting in prolonged decrease in plasma calcium concentration.

Gowen and Mundy (1986) have also shown that γ-interferon blocks the bone-resorbing effects of IL-1 and TNF. γ-Interferon appears to be a more effective inhibitor of bone resorption stimulated by IL-1 and TNF than bone resorption stimulated by PTH or $1,25(OH)_2D_3$. Similarly, Kurihara and Roodman (1990) have shown that other interferons can also inhibit osteoclast formation in human marrow cultures, suggesting that the interferons as a class are inhibitors of bone resorption.

3. *Interleukin-4*

IL-4 is a product of activated T cells with effects on both immunological and hematopoietic processes. Shioi *et al.* (1991) reported that IL-4 inhibited the formation of osteoclasts from murine bone marrow cells cocultured with stromal cells, and Watanabe *et al.* (1990) showed that IL-4 inhibited bone resorption in organ cultures. Others have reported similar results (Riancho *et al.*, 1993a).

Nakano *et al.* (1994) examined the *in vivo* effects of IL-4 on spontaneous and stimulated mouse osteoclast formation. EC-GI cells, which produce PTHrP and IL-1, were explanted into nude mice. After the mice became hypercalcemic, treatment with a continuous infusion of IL-4 returned the calcium levels to normal. Histomorphometric analysis revealed that IL-4 inhibited osteoclast formation in these mice, with a decrease in osteoclastic surface and in the number of osteoclasts per normal bone surface. Furthermore, transgenic mice overexpressing IL-4 develop an osteopenic syndrome that may be likened to osteoporosis (Lewis *et al.*, 1993). Thus, IL-4 exerts inhibitory effects on osteoclasts and osteoblasts, both *in vivo* and *in vitro*.

IV. Summary

Systemic hormones and cytokines play important roles in regulating both osteoblast and osteoclast activity. These cytokines can have either positive or negative effects on the growth and differentiation of bone cells. These effects appear to be dependent on the model systems use to assess them, as well as the species tested. In the near future, other autocrine–paracrine factors will be identified that enhance osteoblast and osteoclast activity, and model systems should be available to further delineate their effects on cells in the osteoblast lineage. Use of transgenic mice with genes targeted to the osteoblast and osteoclast may further reveal the mechanisms responsible for the growth and differentiation of these cells, as well as produce immortalized cell lines that more accurately reflect the cell biology of the osteoclast and osteoblast *in vivo*.

REFERENCES

Alsina, M., Boyce, B., Devlin, R., Anderson, J. L., Craig, F., Mundy, G. R., and Roodman, G. D. (1995). Development of an in vivo model of human multiple myeloma bone disease. *Blood,* in press.

Barengolts, E. I., Lathon, P. V., Lindh, F., and Kukreja, S. C. (1994). Cytokines may be responsible for the inhibited bone formation in hypercalcemia of malignancy. *J. Bone Miner. Res.* **9**(Suppl 1), S138.

Bellows, C. G., Aubin, J. E., Heersche, H. M. N., and Antosz, M. E. (1986). Mineralized bone modules formed in vitro from enzymatically released cat calvaria cell populations. *Calcif. Tissue Int.* **38,** 143–154.

Birch, M. A., Ginty, A. F., Walsh, C. A., Fraser, W. D., Gallagher, J. A., and Bilbe, G. (1993). PCR detection of cytokines in normal human and pagetic osteoblast-like cells. *J. Bone Miner. Res.* **8,** 115–1162.

Blair, H. C., Teitelbaum, S. L., Ghiselli, R., and Gluck, S. (1989). Osteoclastic bone resorption by a polarized vacuolar proton pump. *Science* **245,** 855–857.

Body, J. J., Fernandez, G., Lacroix, M., Vandenbussche, P., and Content, J. (1994). Regulation of lymphocyte calcitonin receptors by interleukin-1 and interleukin-6. *Calcif. Tissue Int.* **55,** 109–113.

Boyce, B. F., Aufdemorte, T. B., Garrett, I. R., Yates, A. J. P., and Mundy, G. R. (1989a). Effects of interleukin-1 on bone turnover in normal mice. *Endocrinology* **123,** 1142–1150.

Boyce, B. F., Yates, A. J. P., Mundy, G. R. (1989b). Bolus injections of recombinant human interleukin-1 cause transient hypocalcemia in normal mice. *Endocrinology* **125,** 2780–2783.

Burger, E. H., Van der Meer, J. W. M., Gevel, J. S., Gribnau, J. C., Thesingh, C. W., and van Furth, R. (1982). In vitro formation of osteoclasts from long-term cultures of bone marrow mononuclear phagocytes. *J. Exp. Med.* **156,** 1604–1614.

Canalis, E. (1993). Systemic and local factors and the maintenance of bone quality. *Calcif. Tissue Int.* **53**(Suppl), S90–S93.

Canalis, E. (1994). Editorial: Skeletal growth factors and aging. *J. Clin. Endocrinol. Metab.* **78,** 1009–1010.

Canalis, E., and Gabbitas, B. (1994). Bone morphogenetic protein 2 increases insulin-like growth factor I and II transcripts and polypeptide levels in bone cell cultures. *J. Bone Miner. Res.* **9,** 1999–2005.

Canalis, E., and Gabbitas, B. (1995). Skeletal growth factors regulate the synthesis of insulin-like growth factor binding protein-5 in bone cell cultures. *J. Biol. Chem.* **270,** 10771–10776.

Canalis, E., and Lian, J. B. (1985). 1,25-Dihydroxyvitamin D2 effects on collagen and DNA synthesis in periosteum and periosteum-free calvaria. *Bone* **6,** 457–460.

Canalis, E., Varghese, S., McCarthy, T. L., and Centrella, M. (1992). Role of platelet derived growth factor in bone cell function. *Growth Regul.* **2,** 151–155.

Canalis, E., Pash, J., Gabbitas, B., Rydziel, S., and Varghese, S. (1993). Growth factors regulate the synthesis of insulin-like growth factor-I in bone cell cultures. *Endocrinology* **133,** 33–38.

Centrella, M., and Canalis, E. (1985). Transforming and non-transforming growth factors are present in medium conditioned by fetal rat calvariae. *Proc. Natl. Acad. Sci. U.S.A.* **82,** 7335–7339.

Centrella, M., McCarthy, T. L., Kusmik, W. F., and Canalis, E. (1992). Isoform-specific regulation of platelet-derived growth factor activity and binding in osteoblast-enriched cultures from fetal rat bone. *J. Clin. Invest.* **89,** 1076–1084.

Centrella, M., Horowitz, M. C., Wozney, J. M., and McCarthy, T. L. (1994). Transforming growth factor-β gene family members and bone. *Endocr. Rev.* **15,** 27–39.

Chambers T. J., McSheehy, P. M. J., Thomson, B. M., and Fuller, K. (1985). The effect of calcium-regulating hormones, prostaglandins on bone resorption by osteoclasts disaggregated from neonatal rabbit bones. *Endocrinology* **116,** 234–239.

Chaudhary, L. R., Spelsberg, T. C., and Riggs, B. L. (1992). Production of various cytokines by normal human osteoblast-like cells in response to interleukin-1 beta and tumor necrosis factor-alpha: Lack of regulation by 17 beta-estradiol. *Endocrinology* **130,** 2528–2534.

Chen, T. L., Cone, C. M., and Morey-Hilton, E. (1983). 1,25-Dihydroxyvitamin D3 receptors in cultured rat osteoblast-like cells. *J. Biol. Chem.* **258,** 4350–4355.

Chenu, C., Pfeilschifter, J., Mundy, G. R., and Roodman, G. D. (1988). Transforming growth factor beta inhibits formation of osteoclast-like cells in long-term human marrow cultures. *Proc. Natl. Acad. Sci. U.S.A.* **85,** 5683–5687.

Chenu, C., Kurihara, N., Mundy, G. R., and Roodman, G. D. (1990). Prostaglandin E_2 inhibits formation of osteoclast-like cells in long-term human marrow cultures but is not a mediator of the inhibitory effects of transforming growth factor-β. *J. Bone Miner. Res.* **5,** 677–681.

Chyun, Y. S., and Raisz, L. G. (1984). Stimulation of bone formation by prostaglandin E2. *Prostaglandins* **27,** 97–103.

Chyun, Y. S., Kream, B. E., and Raisz, L. G. (1984). Cortisol decreases bone formation by inhibiting periosteal cell proliferation. *Endocrinology* **114,** 4777–4780.

Coccia, P. F., Krivit, W., Cervenka, J., Lawson, C., Kersey, J. H., Kim, T. H., Nesbit, M. E., Ramsay, N. K. C., Warketin, P. I., Teitelbaum, S. L., Kahn, A. J., and Brown, D. M. (1980). Successful bone marrow transplantation for infantile malignant osteopetrosis. *New Engl. J. Med.* **302,** 701–708.

Cozzolino, F., Torcia, M., Aldinucci, D., Rubartelli, A., Miliani, A., Shaw, A. R., Lansdor, P. M., and Diguglielmo, R. (1989). Production of interleukin-1 by bone marrow myeloma cells. *Blood* **74,** 380–387.

Cunningham, N. S., Paralkar, V., and Reddi, A. H. (1992). Osteogenin and recombinant bone morphogenetic protein 2B are chemotactic for human monocytes and stimulate transforming growth factor beta 1 mRNA expression. *Proc. Natl. Acad. Sci. U.S.A.* **89,** 11740–11744.

De La Mata, J., Uy, H. L., Guise, T. A., Story, B., Boyce, B. F., Mundy, G. F., and Roodman, G. D. (1995). IL-6 enhances hypercalcemia and bone resorption mediated by PTH-rP in vivo. *J. Clin. Invest.* **95,** 2846–2852.

Delany, A. M., Dong, Y., and Canalis, E. (1994a). Mechanisms of glucocorticoid action in bone cells. *J. Cell. Biochem.* **56,** 295–302.

Delany, A. M., Pash, J. M., and Canalis, E. (1994b). Cellular and clinical perspectives on skeletal insulin-like growth factor I. *J. Cell. Biochem.* **55,** 328–333.

Dempster, D. W., Cosman, F., Parisien, M., Shen, V., and Lindsay, R. (1993). Anabolic actions of parathyroid hormone on bone. *Endoc. Rev.* **14,** 690–709.

Devlin, R. D., Bone, H. G., and Roodman, G. D. (1995a). Interleukin-6: A potential mediator of the massive osteolysis in patients with Gorham-Stout disease. *J. Clin. Endocrinol. Metab.*, submitted.

Devlin, R. D., Reddy, S. V., and Roodman, G. D. (1995b). Annexin II increases osteoclast formation by stimulating the proliferation of osteoclast precursors in human marrow cultures. *Endocrinology,* submitted.

Dietrich, J. W., Canalis, E. M., and Maina, D. M. (1976a). Hormonal control of bone collagen synthesis in vitro: Effects of parathyroid hormone and calcitonin. *Endocrinology* **98,** 943–949.

Dietrich, J. W., Canalis, E. M., and Maina, D. M. (1976b). Dual effects of glucocorticoids on bone collagen synthesis. *Pharmacology* **18,** 234–240.

Dieudonne, S. C., Foo, P., van Zoelen, E. J., and Burger, E. H. (1991). Inhibiting and stimulating effects of TGF-beta 1 on osteoclastic bone resorption in fetal mouse bone organ cultures. *J. Bone Miner. Res.* **6,** 479–487.

Dong, Y., and Canalis, E. (1995). Insulin-like growth factor (IGF) I and retinoic acid

induce the synthesis of IGF-binding protein 5 in rat osteoblastic cells. *Endocrinology* **136**, 2000–2006.
Dunstan, C. R., Garrett, I. R., Adams, R., Burgess, W., Jaye, M., Yongs, T., Boyce, R., and Mundy, G. R. (1995). Systemic fibroblast growth factor (FGF-1) prevents bone loss, increases new bone formation and restores trabecular microarchitecture in ovariectomized rats. *J. Bone Miner. Res.* **10**(Suppl 1), S198.
Eisenberg, S. P., Evans, R. J., Arend, W. P., Verderber, E., Brewer, M. T., Hannum, C. H., and Thompson, R. C. (1990). Primary structure and functional expression from complementary DNA of a human interleukin-1 receptor antagonist. *Nature* (London) **343**, 341–346.
Elias, J. A., Tang, W., and Horowitz, M. C. (1995). Cytokine and hormonal stimulation of human osteosarcoma interleukin-11 production. *Endocrinology* **136**, 489–498.
Eriksen, E. F., Colvard, D. S., Berg, N. J., Graham, M. L., Mann, K. G., Spelsberg, T. C., and Riggs, B. L. (1988). Evidence of estrogen receptors in normal human osteoblastlike cells. *Science* **241**, 84–86.
Ernst, M., Heath, J. K., and Rodan, G. A. (1989). Estradiol effects on proliferation, messenger ribonucleic acid for collagen and insulin-like growth factor-I and parathyroid hormone-stimulated adenylate cyclase activity in osteoblastic cells from calvariae and long bones. *Endocrinology* **125**, 825–833.
Feyen, J. H., Elford, P., Di Padova, F. E., and Trechsel, U. (1989). Interleukin-6 is produced by bone and modulated by parathyroid hormone. *J. Bone Miner. Res.* **4**, 633–638.
Fried, R. M., Voelkel, E. F., Rice, R. H., Levine, L., Gaffney, E. V., and Tashjian, A. H. (1989). Two squamous cell carcinomas not associated with humoral hypercalcemia produce a potent bone resorption-stimulating factor which is interleukin-1 alpha. *Endocrinology* **125**, 742–751.
Gabbitas, B., and Canalis, E. (1995). Bone morphogenetic protein-2 inhibits the synthesis of insulin-like growth factor-binding protein-5 in bone cell cultures. *Endocrinology* **136**, 2397–2403.
Gallwitz, W. E., Mundy, G. R., Lee, C. H., Qiao, M., Roodman, G. D., Raftery, M., Gaskell, S. J., and Bonewald, L. F. (1993). 5-Lipoxygenase metabolites of arachidonic acid stimulate isolated osteoclasts to resorb calcified matrices. *J. Biol. Chem.* **268**, 10087–10094.
Garrett, I. R., Durie, B. G. M., and Nedwin, G. E. (1987). Production of the bone-resorbing cytokine lymphotoxin by cultured human myeloma cells. *New Engl. J. Med.* **317**, 526–532.
Ghosh-Choudhuury, N., Harris, M. A., Feng, J. Q., Mundy, G. R., and Harris, S. E. (1994). Expression of the BMP 2 gene during bone cell differentiation. *Crit. Rev. Eukaryotic Gene Expression* **4**, 345–355.
Gilardetti, R. S., Chaibi, M. S., Stroumza, J., Williams, S. R., Antoniades, H. N., Carnes, D. C., and Graves, D. T. (1991). High-affinity binding of PDGF-AA and PDGF-BB to normal human osteoblastic cells and modulation by interleukin-1. *Am. J. Physiol.* **261**, C980–C985.
Girasole, G., Jilka, R. L., Passeri, G., Boswell, S., Boder, G., Williams, D. C., and Manolagas, S. C. (1992). 17 Beta-estradiol inhibits interleukin-6 production by bone marrow-derived stromal cells and osteoblasts in vitro: A potential mechanism for the antiosteoporotic effect of estrogens. *J. Clin. Invest.* **89**, 883–891.
Gowen, M., and Mundy, G. R. (1986). Actions of recombinant interleukin-1, interleukin-2, and interferon gamma on bone resorption in vitro. *J. Immunol.* **136**, 2478–2482.

Gowen, M., Meikle, M. C., and Reynolds, J. J. (1983). Stimulation of bone resorption in vitro by a non-prostanoid factor released by human monocytes in culture. *Biochim. Biophys. Acta* **762**, 471–474.

Graves, D. T., and Owen, A. J. (1983). Evidence that a human osteosarcoma cell line which secretes a mitogen similar to platelet-derived growth factor requires growth factors present in platelet-poor plasma. *Cancer Res.* **43**, 83–87.

Graves, D. T., Owen, A. J., and Barth, R. K. (1984). Detection of c-cis transcripts and synthesis of PDGF-like proteins by human osteosarcoma cells. *Science* **226**, 972–974.

Gray, T. K., Flynn, T. C., Gray, K. M., and Nabell, L. M. (1987). 17β-estradiol acts directly on the clonal osteoblastic cell line AMBER. *Proc. Natl. Acad. Sci. U.S.A.* **84**, 6267–6271.

Greenfield, E. M., Gornik, S. A., Horowitz, M. C., Donahue, H. J., and Shaw, S. M. (1993). Regulation of cytokine expression in osteoblasts by parathyroid hormone: Rapid stimulation of interleukin-6 and leukemia inhibitory factor mRNA. *J. Bone Miner. Res.* **8**, 1163–1171.

Hakeda, Y., Nakatani, Y., and Kurihara, N. (1985). Prostaglandin E2 stimulates collagen and noncollagen protein synthesis and prolyl hydroxylase activity in osteoblastic clone MC3T3-E1 cells. *Biochem. Biophys. Res. Commun.* **126**, 340–345.

Hakeda, Y., Hiura, K., Sato, T., Olazaki, R., Matsumoto, T., Ogata, E., Ishitani, R., and Kumegawa, M. (1989). Existence of parathyroid hormone binding sites on murine hemopoietic blast cells. *Biochem. Biophys. Res. Commun.* **163**, 1481–1486.

Hanevold, C. D., Yamaguchi, D. T., and Jordan, S. C. (1993). Tumor necrosis factor alpha modulates parathyroid hormone action in UMR-106-01 osteoblastic cells. *J. Bone Miner. Res.* **8**, 1191–1200.

Hannum, C. H., Wilcox, C. J., and Arend, W. P. (1990). Interleukin-1 receptor antagonist activity of a human interleukin-1 inhibitor. *Nature* (London) **343**, 336–340.

Harris, S. E., Feng, J. Q., Harris, M. A., Ghosh-Choudhury, N., Dallas, M. R., Wozney, J., and Mundy, G. R. (1995). Recombinant bone morphogenetic protein 2 accelerates bone cell differentiation and stimulates BMP-2 mRNA expression and BMP-2 promoter activity in primary fetal rat calvarial osteoblast cultures. *Mol. Cell. Differ.* **3**, 137–155.

Hauschka, P. V., Maurakos, E., and Lafrati, M. D. (1986). Growth factors in bone matrix. *J. Biol. Chem.* **261**, 12665–12674.

Hiraga, T., Nakajima, T., and Ozawa, H. (1995). Bone resorption induced by a metastatic human melanoma cell line. *Bone* **16**, 349–356.

Hock, J. M., and Canalis, E. (1994). Platelet-derived growth factor enhances bone cell replication, but not differentiated function of osteoblasts. *Endocrinology* **134**, 1423–1428.

Hock, J. M., and Fonseca, J. (1990). Anabolic effect of human synthetic parathyroid hormone- (1-34) depends on growth hormone. *Endocrinology* **127**, 1804–1810.

Horowitz, M. C. (1993). Cytokines and estrogen in bone: Anti-osteoporotic effects. *Science* **260**, 626–627.

Hurley, M. M., Abreau, C., Gronowicz, G., Kawaguchi, H., and Lorenzo, J. (1994). Expression and regulation of basic fibroblast growth factor mRNA levels in mouse osteoblastic MC3T3-E1 cells. *J. Biol. Chem.* **269**, 9392–9396.

Ibbotson, K. J., Twardzik, D. R., D'Souza, S. M., Hargreaves, W. R., Todaro, G. D., and Mundy, G. R. (1985). Stimulation of bone resorption in vitro by synthetic transforming growth factor alpha. *Science* **228**, 1007–1009.

Ignotz, R., and Massague, J. (1986). Transforming growth factor-beta stimulates the expression of fibronectin and collagen and their incorporation into the extracellular matrix. *J. Biol. Chem.* **261**, 4337–4445.

Ignotz, R., Endo, T., and Massague, J. (1987). Regulation of fibronectin and type 1 collagen mRNA levels by transforming growth factor-beta. *J. Biol. Chem.* **262**, 6443–6446.

Ishibashi, H., Karube, S., Yamakawa, A., and Koshihara, Y. (1995). Interleukin-4 stimulates pro-alpha 1 (VI) collagen gene expression in cultured human osteoblast-like cells. *Biochem. Biophys. Res. Commun.* **211**, 727–734.

Jilka, R. L., Hangoc, G., Girasole, G., Passeri, G., Williams, D. C., Abrams, J. S., Boyce, B., Broxmeyer, H., and Manolagas, S. C. (1992). Increased osteoclast development after estrogen loss: Mediation by interleukin-6. *Science* **257**, 88–91.

Karaplis, A. C., Luz, A., Glowacki, J., Bronson, R. T., Tybulewicz, V. L., Kronenberg, H. M., and Mulligan, R. C. (1994). Lethal skeletal dysplasia from targeted disruption of the parathyroid hormone-related peptide genes. *Genes Dev.* **8**, 277–289.

Katz, M. S., Gutierrez, G. E., Mundy, G. R., Hymer, T. K., Caulfield, M. P., and McKee, R. L. (1992). Tumor necrosis factor and interleukin 1 inhibit parathyroid hormone-responsive adenylate cyclase in clonal osteoblast-like cells by down-regulating parathyroid hormone receptors. *J. Cell Physiol.* **153**, 206–213.

Kawano, M., Tanaka, H., Ishikawa, H., Nobuyoshi, M., Iwato, K., Asaoku, H., Tanabe, O., and Kuramoto, A. (1989). Interleukin-1 accelerates autocrine growth of myeloma cells through interleukin-6 in human myeloma. *Blood* **73**, 2145–2148.

Keeting, P. E., Scott, R. E., Colvard, D. S., Han, I. K., Spelsberg, T. C., and Riggs, B. L. (1991). Lack of a direct effect of estrogen on proliferation and differentiation of normal human osteoblast-like cells. *J. Bone Miner. Res.* **6**, 297–304.

Kimble, R. B., Kitazawa, R., Vannice, J. L., and Pacifici, R. (1994). Persistent bone-sparing effect of interleukin-1 receptor antagonist: A hypothesis on the role of IL-1 in ovariectomy-induced bone loss. *Calcif. Tissue Int.* **55**, 260–265.

Kitazawa, R., Kimble, R. B., Vannice, J. L., Kung, V. T., and Pacifici, R. (1994). Interleukin-1 receptor antagonist and tumor necrosis factor binding protein decrease osteoclast formation and bone resorption in ovariectomized mice. *J. Clin. Invest.* **94**, 2397–2406.

Komm, B. S., Terpening, C. M., Benz, D. J., Graeme, K. A., Omalley, B. W., and Haussler, M. R. (1988). Estrogen binding receptor mRNA, and biologic response in osteoblast-like osteosarcoma cells. *Science* **241**, 81–84.

Kurihara, N., and Roodman, G. D. (1990). Interferons-α and -γ inhibit interleukin-1β-stimulated osteoclast-like cell formation in long-term human marrow cultures. *J. Interferon Res.* **10**, 541–547.

Kurihara, N., Bertolini, D., Suda, T., Akiyama, Y., and Roodman, G. D. (1990a). IL-6 stimulates osteoclast-like multinucleated cell formation in long-term human marrow cultures by inducing IL-1 release. *J. Immunol.* **144**, 4226–4230.

Kurihara, N., Chenu, C., Miller, M., Civin, C. I., and Roodman, G. D. (1990b). Identification of committed mononuclear precursors for osteoclast-like cells formed in long-term marrow cultures. *Endocrinology* **126**, 2733–2741.

Lewis, D. B., Liggitt, H. D., Effmann, E. L., Motley, S. T., Teitelbaum, S. L., Jepsen, K. J., Goldstein, S. A., Bonadia, J., Carpenter, J., and Perlmutter, R. M. (1993). Osteoporosis induced in mice by overproduction of interleukin 4. *Proc. Natl. Acad. Sci. U.S.A.* **90**, 11618–11622.

Lind, M., Deleuran, B., Yssel, H., Fink-Eriksen,, E., and Thestrup-Pedersen, K. (1995). IL-4 and IL-13, but not IL-10, are chemotactic factors for human osteoblasts. *Cytokine* **7**, 78–82.

Littlewood, A. J., Russell, J., Harvey, G. R., Hughes, D. E., and Russell, R. G., and Gowen, M. (1991). The modulation of the expression of IL-6 and its receptor in human osteoblasts in vitro. *Endocrinology* **129,** 1513–1520.

Liu, J. P., Baker, J., Perkins, A. S., Robertson, E. J., and Efstratiadis, A. (1993). Mice carrying null mutations of the genes encoding insulin-like growth factor I (lgf-1) and type 1 IGF receptor (lgf1r). *Cell* **75,** 59–72.

Lorenzo, J. A., Quinton, J., Sousa, S., and Raisz, L. G. (1986). Effects of DNA and prostaglandin synthesis inhibitors on the stimulation of bone resorption by epidermal growth factor in fetal rat long-bone cultures. *J. Clin. Invest.* **77,** 1897–1902.

Lorenzo, J. A., Sousa, S. L., Fonseca, J. M., Hock, J. M., and Medlock, E. S. (1987). Colony-stimulating factors regulate the development of multinucleated osteoclasts from recently replicated cells in vitro. *J. Clin. Invest.* **80,** 160–164.

Lowik, C. W. G. M., van der Pluijm, G., Bloys, H., Hoekman, K., Bijvoet, O. L. M., Aarden, L. A., and Papapoulos, S. E. (1989). Parathyroid hormone (PTH) and PTH-like protein (PLP) stimulate interleukin-6 production by osteogenic cells: A possible role of interleukin-6 in osteoclastogenesis. *Biochem. Biophys. Res. Commun.* **162,** 1546–1552.

Lukert, B. P., and Raisz, L. G. (1990). Glucocorticoid-induced osteoporosis: Pathogenesis and management. *Ann. Intern. Med.* **112,** 931–939.

Lyons, R. M., Keski-Oja, J., Moses, H. L. (1988). Proteolytic activation of latent transforming growth factor-β from fibroblast-conditioned medium. *J. Cell Biol.* **106,** 1659–1665.

MacDonald, B. R., Mundy, G. R., Clark, S., Wang, E. A., Kuehl, T. J., Stanley, E. R., and Roodman, G. D. (1986). Effects of human recombinant CSF-GM and highly purified CSF-1 on the formation of multinucleated cells with osteoclast characteristics in long-term bone marrow cultures. *J. Bone Miner. Res.* **1,** 227–233.

MacDonald, B. R., Takahashi, N., McManus, L. M., Holahan, J., Mundy, G. R., and Roodman, G. D. (1987). Formation of multinucleated cells that respond to osteotropic hormones in long-term human bone marrow cultures. *Endocrinology* **120,** 2326–2333.

Mackie, E. J., and Trechsel, U. (1990). Stimulation of bone formation in vivo by transforming growth factor β—remodeling of woven bone and lack of inhibition by indomethacin. *Bone* **11,** 295–300.

Maliakal, J. C., Asahina, I., Hauschka, P. V., and Sampath, T. K. (1994). Osteogenic protein-1 (BMP-7) inhibits cell proliferation and stimulates the expression of markers characteristic of osteoblast phenotype in rat osteosarcoma cells. *Growth Factors* **11,** 227–234.

Manolagas, S. C., and Jilka, R. L. (1995). Bone marrow, cytokines, and bone remodeling: Emerging insights into the pathophysiology of osteoporosis. *New Engl. J. Med.* **332,** 305–311.

Manolagas, S. C., Burton, D. W., and Deftos, L. J. (1981). 1,25-Dihydroxyvitamin D3 stimulates the alkaline phosphatase activity of osteoblast-like cells. *J. Biol. Chem.* **256,** 7115–7117.

Marcelli, C., Yates, A. J. P., and Mundy, G. R. (1990). In vivo effects of human recombinant transforming growth factor beta on bone turnover in normal mice. *J. Bone Miner. Res.* **5,** 1087–1096.

Martin, T. J., Ng, K. W., and Suda, T. (1989). Bone cell physiology. *Endocrinol. Metab. Clin. North Am.* **18,** 833–858.

Mayur, N., Lewis, S., Catherwood, B. D., and Nanes, M. S. (1993). Tumor necrosis factor alpha decreases 1,25-dihydroxyvitamin D3 receptors in osteoblastic ROS 17/2.8 cells. *J. Bone Miner. Res.* **8**, 997–1003.

McCarthy, T. L., Centrella, M., and Canalis, E. (1989). Parathyroid hormone enhances the transcript and polypeptide levels of insulin-like growth factor-I in osteoblast-enriched cultures from fetal rat bone. *Endocrinology* **124**, 1247–1253.

McSheehy, P. H. J., and Chambers, T. J. (1986). Osteoblastic cells mediate osteoclastic responsiveness to parathyroid hormone. *Endocrinology* **118**, 824–828.

Miyazono, K., Hellman, U., and Wernstedt, C. (1988). Latent high molecular weight complex of transforming growth factor $\beta 1$: Purification from human platelets and structural characterization. *J. Biol. Chem.* **263**, 6407–6415.

Morohashi, T., Corboz, V. A., Fleisch, H., Cecchini, M. G., and Felix, R. (1994). Macrophage colony-stimulatnig factor restores bone resorption in op/op bone in vitro in conjunction with parathyroid hormone or 1,25-dihydroxyvitamin D_3. *J. Bone Miner. Res.* **9**, 401–407.

Mundy, G. R. (1991). The effects of TGF-beta on bone. *Ciba Found. Symp.* **157**, 137–143.

Mundy, G. R. (1995a). Osteoblasts, bone formation and mineralization. In "Bone Remodeling and Its Disorders" (I. Fogelman, ed.), pp. 27–38. Martin Dunitz, London.

Mundy, G. R. (1995b). Factors regulating bone resorbing and bone forming cells. In "Bond Remodeling and Its Disorders" (I. Fogelman, ed.), pp. 39–65. Martin Dunitz, London.

Mundy, G. R., and Roodman, G. D. (1987). Osteoclast ontogeny and function. In "Bone & Mineral Research" (W. A. Peck, ed.), pp. 209–280. Elsevier, Amsterdam.

Nakano, Y., Watanabe, K., Morimoto, I., Okada, Y., Ura, K., Sato, K., Kasono, K., Nakamura, T., and Eto, S. (1994). Interleukin-4 inhibits spontaneous and parathyroid hormone-related protein-stimulated osteoclast formation in mice. *J. Bone Miner. Res.* **9**, 1533–1539.

Narbaitz, R., Stumpf, W., and Sar, M. (1983). Autoradiographic demonstration of target cells for 1,25-dihydroxycholecalciferol in bones from fetal rats. *Calcif. Tissue Int.* **35**, 177–182.

Nesbitt, S. A., and Horton, M. A. (1995). Osteoclast annexins bind collagen and play a role in bone resorption. *J. Bone Miner. Res.* **10**(Suppl 1), S221.

Noda, M. (1989). Transcriptional regulation of osteocalcin production by transforming growth factor-β in rat osteoblast-like cells. *Endocrinology* **124**, 612–617.

Noda, M., and Camilliere, J. J. (1989). In vivo stimulation of bone formation by transforming growth factor-β. *Endocrinology* **124**, 2991–2994.

Noda, M., and Rodan, G. A. (1986). Type β transforming growth factor inhibits proliferation and expression of alkaline phosphatase in murine osteoblast-like cells. *Biochem. Biophys. Res. Commun.* **140**, 56–60.

Noda, M., and Rodan, G. A. (1987). Type β transforming growth factor (TGF-β) regulation of alkaline phosphatase expression and other phenotype-related mRNAs in osteoblastic rat osteosarcoma cells. *J. Cell Physiol.* **133**, 426–437.

Noda, M., Kato, I., Kirayama, T., and Matsuda, F. (1982). Mode of action of staphylococcal leukocidin: Effects of the S and F components on the activities of membrane-associated enzymes of rabbit polymorphonuclear leukocytes. *Infect. Immun.* **35**, 38–45.

Noda, M., Yoon, K., and Prince, C. W. (1988). Transcriptional regulation of osteopontin production in rat osteosarcoma cells by type β transforming growth factor. *J. Biol. Chem.* **263**, 13916–13921.

Oursler, M. J. (1994). Osteoclast synthesis, secretion and activation of latent transforming growth factor beta. *J. Bone Miner. Res.* **9**, 443–452.

Oursler, M. J., Bell, L. V., Clevinger, B., and Osdoby, P. (1985). Identification of osteoclast specific monoclonal antibodies. *J. Cell Biol.* **100**, 1592–1600.

Oursler, M. J., Cortese, C., Keeting, P. E., Anderson, M. A., Bonde, S. K., Riggs, B. L., and Spelsberg, T. C. (1991a). Modulation of transforming growth factor-β production in normal human osteoblast-like cells by 17β-estradiol and parathyroid hormone. *Endocrinology* **129**, 3313–3320.

Oursler, M., Osdoby, P., Pyfferoen, J., Riggs, B. L., and Spelsberg, T. C. (1991b). Avian osteoclasts as estrogen target cells. *Proc. Natl. Acad. Sci. U.S.A.* **88**, 6613–6617.

Pacifici, R., Vannice, J. L., Rifas, L., and Kimble, R. B. (1993). Monocytic secretion of interleukin-1 receptor antagonist in normal and osteoporotic women: Effects of menopause and estrogen/progesterone therapy. *J. Clin. Endocrinol. Metab.* **77**, 1135–1141.

Panagakos, F. S., Hinojosa, L. P., and Kumar, S. (1994). Formation and mineralization of extracellular matrix secreted by an immortal human osteoblastic cell line: Modulation by tumor necrosis factor-alpha. *Inflammation* **18**, 267–284.

Pash, J. M., Delany, A. M., Adamo, M. L., Roberts, C. T., LeRoith, D., and Canalis, E. C. (1995). Regulation of insulin-like growth factor I transcription by prostaglandin E_2 in osteoblast cells. *Endocrinology* **136**, 33–88.

Pfeilschifer, J., and Mundy, G. R. (1987). Modulation of transforming growth factor β activity in bone cultures by osteotropic hormones. *Proc. Natl. Acad. Sci. U.S.A.* **84**, 2024–2028.

Pfeilschifter, J., D'Souza, S. M., and Mundy, G. R. (1987). Effects of transforming growth factor-β on osteoblastic osteosarcoma cells. *Endocrinology* **121**, 212–218.

Pfeilschifter, J., Chenu, C., Bird, A., Mundy, G. R., and Roodman, G. D. (1989a). Interleukin-1 and tumor necrosis factor stimulate the formation of human osteoclast-like cells in vitro. *J. Bone Miner. Res.* **4**, 113–118.

Pfeilschifter, J., Wolf, O., and Naumann, A. (1989b). Chemotactic response of osteoblast-like cells to transforming growth factor β. *J. Bone Miner. Res.* **5**, 825–830.

Price, P. A., and Baukol, S. A. (1980). 1,25-Dihydroxyvitamin D3 increases synthesis of the vitamin K-dependent bone protein by osteosarcoma cells. *J. Biol. Chem.* **255**, 11660–11663.

Raisz, L. G., Kream, B. E., and Smith, M. D. (1980). Comparison of the effects of vitamin D metabolites on collagen synthesis and resorption of fetal rat bone in organ culture. *Calcif. Tissue Int.* **32**, 135–138.

Reddy, S. V., Takahashi, S., Dallas, M., Williams, R. E., Neckers, L., and Roodman, G. D. (1994). IL-6 antisense deoxyoligonucleotides inhibit bone resorption by giant cells from human giant cell tumors of bone. *J. Bone Miner. Res.* **9**, 753–757.

Riancho, J. A., Zarrabeitia, M. T., and Gonzalez-Macias, J. (1993a). Interleukin-4 modulates osteoclast differentiation and inhibits the formation of resorption pits in mouse osteoclast cultures. *Biochem. Biophys. Res. Commun.* **196**, 678–685.

Riancho, J. A., Zarrabeitia, M. T., Olmos, J. M., Amado, J. A., and Gonzalez-Macias, J. (1993b). Effect of interleukin-4 on human osteoblast-like cells. *Bone Miner.* **21**, 53–61.

Riancho, J. A., Gonzalez-Macias, J., Amado, J. A., Olmos, J. M., and Fernandez-Luna, J. (1995). Interleukin-4 as a bone regulatory factor: Effects on murine osteoblast-like cells. *J. Endocrinol. Invest.* **18**, 174–179.

Rickard, D., Russell, G., and Gowen, M. (1992). Oestadiol inhibits the release of tumour necrosis factor but not interleukin 6 from adult human osteoblasts in vitro. *Osteoporosis Int.* **2**, 94–102.

Robey, P. G. (1989). The biochemistry of bone. *Endocrinol. Metab. Clin. North Am.* **18**, 859–902.

Robey, P. G., Young, M. F., and Flanders, K. C. (1987). Osteoblasts synthesize and respond to transforming growth factor type-β in vitro. *J. Cell Biol.* **105**, 457–463.

Rodan, G. A., and Martin, T. J. (1981). Role of osteoblasts in hormonal control of bone resorption: A hypothesis. *Calcif. Tissue Int.* **33**, 349–351.

Roodman, G. D., Kurihara, N., Ohsaki, Y., Kukita, A., Hosking, D., Demulder, A., and Singer, F. R. (1991). Interleukin-6: A potential autocrine/paracrine factor in Paget's disease of bone. *J. Clin. Invest.* **89**, 46–52.

Rouleau, M. F., Mitchell, J., and Goltzman, D. (1988). In vivo distribution of parathyroid hormone receptors in bone: Evidence that a predominant osseous target cell is not the mature osteoblast. *Endocrinology* **123**, 187–191.

Rouleau, M. F., Mitchell, J., and Goltzman, D. (1990). Characterization of the major parathyroid hormone target cell in the endosteal metaphysics of rat long bones. *J. Bone Miner. Res.* **5**, 1043–1053.

Rusinko, R., Yin, J. J., Yee, J., Saad, T., Mundy, G. R., and Guise, T. A. (1995). Parathyroid hormone (PTH) excess is associated with increased interleukin-6 and soluble IL-6 receptor (sILI-6r) production. *J. Bone Miner. Res.* **10**(Suppl 1), S500.

Sato, K., Kasono, K., and Fuji, Y. (1987). Tumor necrosis factor type α stimulates mouse osteoblast-like cells (MC3T3-E1) to produce macrophage-colony stimulating activity and prostaglandin E2. *Biochem. Biophys. Res. Commun.* **145**, 323–329.

Sato, K., Fujii, Y., Kasono, K., Ozawa, M., Imamura, H., Kanaji, K., Kurosawa, H., Tsushima, T., and Shizume, K. (1989). Parathyroid hormone-related protein and interleukin-1α synergistically stimulate bone resorption in vitro and increase the serum calcium concentration in mice in vivo. *Endocrinology* **124**, 2172–2178.

Schneider, H. G., Allan, E. H., Moseley, J. M., Martin, T. J., and Findlay, D. M. (1991). Specific down-regulation of parathyroid hormone (PTH) receptors and responses to PTH by tumour necrosis factor alpha and retinoic acid in UMR-106-06 osteoblast-like osteosarcoma cells. *Biochem. J.* **280**, 451–457.

Shioi, A., Teitelbaum, S. L., Ross, F. P., Welgus, H. G., Suzuki, H., Ohara, J., and Lacey, D. L. (1991). Interleukin-4 inhibits murine osteoclast formation in vitro. *J. Cell Biol.* **47**, 272–277.

Shuto, T., Kukita, T., Hirata, M., Jimi, E., and Koga, T. (1994). Dexamethasone stimulates osteoclast-like cell formation by inhibiting granulocyte-macrophage colony-stimulating factor production in mouse bone marrow cultures. *Endocrinology* **134**, 1121–1126.

Smith, E. P., Boyd, J., Frank, G. R., Takahashi, H., Cohen, R. M., Specker, B., Williams, T. C., Lubahn, D. B., and Korach, K. S. (1994). Estrogen resistance caused by a mutation in the estrogen-receptor gene in a man. *New Engl. J. Med.* **331**, 1088–1089.

Sorell, M., Kapoor, N., Kirkpatrick, D., Rosen, J., Chaganti, R., Lopez, C., Dupont, B., Pollack, M., Terrin, B., Harris, M., Vine, D., Rose, J., Goossen, C., Lane, J., Good, R., and O'Reilly, R. J. (1981). Marrow transplantation for juvenile osteopetrosis. *Am. J. Med.* **70**, 1280–1287.

Spencer, E. M., Tokunaga, A., and Hunt, T. K. (1993). Insulin-like growth factor binding protein-3 is present in the alpha-granules of platelets. *Endocrinology* **132**, 996–1001.

Sporn, M. B., Roberts, A. B., and Wakefield, L. M. (1987). Some recent advances in the chemistry and biology of transforming growth factor-β. *J. Cell Biol.* **105,** 1039–1045.

Stein, G. S., and Lian, J. B. (1993). Molecular mechanism mediating proliferation/differentiation interrelationships during progressive development of the osteoblast phenotype. *Endoc. Rev.* **14,** 424–442.

Subramaniam, M., Oursler, M. J., Rasmussen, K., Riggs, B. L., and Spelsberg, T. C. (1995). TGF-beta regulation of nuclear proto-oncogenes and TGF-beta gene expression in normal human osteoblast-like cells. *J. Cell. Biochem.* **57,** 62–61.

Taichman, R. S., and Hauschka, P. V. (1992). Effects of interleukin-1 beta and tumor necrosis factor-alpha on osteoblastic expression of osteocalcin and mineralized extracellular matrix in vitro. *Inflammation* **16,** 587–601.

Takahashi, N., MacDonald, B. R., Hon, J., Winkler, M. E., Derynck, R., Mundy, G. R., and Roodman, G. D. (1986a). Recombinant human transforming growth factor-α stimulates the formation of osteoclast-like cells in long-term human marrow cultures. *J. Clin. Invest.* **78,** 894–898.

Takahashi, N., Mundy, G. R., and Roodman, G. D. (1986b). Recombinant human gamma interferon inhibits formation of osteoclast-like cells by inhibiting fusion of their precursors. *J. Immunol.* **137,** 3544–3549.

Takahashi, N., Mundy, G. R., Kuehl, T. J., and Roodman, G. D. (1987). Osteoclast-like cell formation in fetal and newborn long-term baboon marrow cultures is more sensitive to 1,25-dihydroxyvitamin D_3 than adult long-term marrow cultures. *J. Bone Miner. Res.* **2,** 311–317.

Takahashi, N., Yamana, H., Yoshiki, S., Roodman, G. D., Mundy, G. R., Jones, S. J., Boyde, A., and Suda, T. (1988). Osteoclast-like cell formation and its regulation by osteotropic hormones in mouse bone marrow cultures. *Endocrinology* **122,** 1373–1382.

Takahashi, N., Udagawa, N., Akatsu, T., Tanaka, H., Shionome, M., and Suda, T. 1991). Role of colony-stimulating factors in osteoclast development. *J. Bone Miner. Res.* **6,** 977–985.

Takahashi, S., Reddy, S. V., Chirgwin, J. M., Devlin, R., Haipek, C., Anderson, J., and Roodman, G. D. (1984). Cloning and identification of Annexin II as an autocrine/paracrine factor that increases osteoclast formation and bone resorption. *J. Biol. Chem.* **269,** 28696–28701.

Takahashi, S., Goldring, S., Katz, M., Hilsenbeck, S., Williams, R., and Roodman, G. D. (1985a). Downregulation of calcitonin receptor mRNA expression by calcitonin during human osteoclast-like cell differentiation. *J. Clin. Invest.* **95,** 167–171.

Takahashi, S., Reddy, S. V., Dallas, M., Devlin, R., Chou, J. Y., and Roodman, G. D. (1995b). Development and characterization of a human marrow stromal cell line that enhances osteoclast-like cell formation. *Endocrinology* **136,** 1441–1449.

Tashjian, A. H., Voelkel, E. F., Lazzaro, M., Goad, D., Bosma, T., and Levine, L. (1985). Alpha and beta transforming growth factors stimulate prostaglandin production and bone resorption in cultured mouse calvaria. *Proc. Natl. Acad. Sci. U.S.A.* **82,** 4535–4538.

Teti, L. T., Rizzoli, R., and Zambonin, Z. A. (1991). Parathyroid hormone binding to cultured avian osteoclasts. *Biochem. Biophys. Res. Commun.* **174,** 1217–1222.

Thomson, B. M., Saklatvala, J., and Chambers, T. J. (1986). Osteoblasts mediate interleukin-1 stimulation of bone resorption by rat osteoclasts. *J. Exp. Med.* **164,** 104–112.

Thomson, B. M., Mundy, G. R., and Chambers, T. J. (1987). Tumor necrosis factors alpha

and beta induce osteoclastic cells to stimulate osteoclastic bone resorption. *J. Immun.* **138**, 775–779.
Tohkin, M., Kakudo, S., Kasai, H., and Arita, H. (1994). Comparative study of inhibitory effects by murine interferon gamma and a new bisphosphonate (alendronate) in hypercalcemic, nude mice bearing human tumor (LJC-1-JCK). *Cancer Immunol. Immunother.* **39**, 155–160.
Turner, R. T., Riggs, B. L., and Spelsberg, T. C. (1994). Skeletal effects of estrogen. *Endoc. Rev.* **15**, 275–300.
Ueno, K., Haba, T., and Woodbury, D. (1985). The effects of prostaglandin E2 in rapidly growing rats: Depressed longitudinal and radial growth and increased metaphyseal hard tissue mass. *Bone,* **6**, 79–86.
Urist, M. R. (1965). Bone: Formation by autoinduction. *Science* **150**, 893–989.
Uy, H. L., Dallas, M., Calland, J. W., Boyce, B. F., Mundy, G. R., and Roodman, G. D. (1995a). Use of an in vivo model to determine the effects of interleukin-1 on cells at different stages in the osteoclast lineage. *J. Bone Miner. Res.* **10**, 295–301.
Uy, H. L., Guise, T. A., De La Mata, J., Taylor, S. D., Story, B. M., Dallas, M. R., Boyce, B. F., Mundy, G. R., and Roodman, G. D. (1995b). Effects of PTHrP and PTH on osteoclasts and osteoclast precursors in vivo. *Endocrinology* **36**, 3207–3212.
Van der Plas, A., Feyen, J. H. M., and Nijweide, P. J. (1985). Direct effect of parathyroid hormone on the proliferation of osteoblast-like cells: A possible involvement of cyclic AMP. *Biochem. Biophys. Res. Commun.* **129**, 918–925.
Walker, D. G. (1972). Congenital osteopetrosis in mice cured by parabiotic union with normal siblings. *Endocrinology* **91**, 916–920.
Walker, D. G. (1973). Osteopetrosis cured by temporary parabiosis. *Science* **180**, 875–880.
Walker, D. G. (1975a). Control of bone resorption by hematopoietic tissue: The induction and reversal of congenital osteopetrosis in mice through the use of bone marrow and splenic transplants. *J. Exp. Med.* **142**, 651–663.
Walker, D. G. (1975b). Bone resorption restored in osteopetrotic mice by transplants of normal bone marrow and spleen cells. *Science* **190**, 784–785.
Walsh, C. A., Birch, M. A., Fraser, W. D., Lawton, R., Dorgan, J., Walsh, S., Sansom, D., Beresford, J. N., and Gallagher, J. A. (1995). Expression and secretion of parathyroid hormone-related protein by human bone-derived cells in vitro: Effects of glucocorticoids. *J. Bone Miner. Res.* **10**, 17–25.
Watanabe, K., Tanaka, Y., Morimoto, I., Yahata, K., Zeki, K., Fugihira, T., Yamashita, U., and Eto, S. (1990). Interleukin-4 as a potent inhibitor of bone resorption. *Biochem. Biophys. Res. Commun.* **172**, 1035–1041.
Watts, C. K. W., Parker, M. G., and King, R. J. B. (1989). Stable transfection of the oestrogen receptor gene into a human osteosarcoma cell line. *J. Steroid Biochem.* **34**, 483–490.
Weir, E., Philbrick, W., Neff, L., Amling, M., Baron, R., and Broadus, A. (1995). Targeted overexpression of parathyroid hormone-related peptide in chondrocytes causes skeletal dysplasia and delayed osteogenesis. *J. Bone Miner. Res.* **10**(Suppl 1), S157.
Weisz, A., and Rosales, R. (1990). Identification of an estrogen response element upstream of the human c-fos gene that binds the estrogen receptor and the AP-1 transcription factor. *Nucleic Acids Res.* **18**, 5097–5106.
Xie, J. F., Stroumza, J., and Graves, D. T. (1994). IL-1 down-regulates platelet-derived growth facactor-alpha receptor gene expression at the transcriptional level in human osteoblastic cells. *J. Immunol.* **153**, 378–383.

Yates, A. J. P., Favarato, G., Aufdemorte, T. B., Marcelli, C., Kester, M. B., Walker, R., Langton, B. C., Bonewald, L., and Mundy, G. R. (1992). Expression of human transforming growth factor α by Chinese hamster ovarian tumors in nude mice causes hypercalcemia and increased osteoclastic bone resorption. *J. Bone Miner. Res.* **7,** 847–853.

Yeh, Y. L., Kang, Y. M., Chaibi, M. S., Xie, J. F., and Graves, D. T. (1993). IL-1 and transforming growth factor-beta inhibit platelet-derived growth factor-AA binding to osteoblastic cells by reducing platelet-derived growth factor-alpha receptor expression. *J. Immunol.* **150,** 5625–5631.

The Molecular Pharmacology of Ovarian Steroid Receptors

ELISABETTA VEGETO,* BRANDEE L. WAGNER,† MARKUS O. IMHOF,† AND DONALD P. MCDONNELL†

Milano Molecular Pharmacology Laboratory, Institute of Pharmacological Science, University of Milan, 20130 Milan, Italy, and †Department of Pharmacology, Duke University Medical Center, Durham, North Carolina 27710

I. Introduction
II. The Mechanism of Action of Estrogen and Progesterone
    A. Estrogen and Progesterone Receptors
    B. Transcriptional Regulation by Estrogen and Progesterone Receptors
    C. Initiation of the Progesterone Receptor-Mediated Signal Transduction Pathway
    D. Initiation of the Estrogen Receptor-Mediated Signal Transduction Pathway
    E. Steroid Hormone Receptors as Transcription Factors
III. Steroid Hormone Receptor Antagonists
    A. Insights into Antihormone Action from Other Systems
    B. The Molecular Mechanism of Action of Antihormones
    C. Antagonism of Progesterone Receptor Function
    D. Antagonism of Estrogen Receptor Function
IV. Final Comments
References

I. INTRODUCTION

Hormones are chemical messengers that are synthesized and secreted by specific cell types into the circulatory system, linking a central point of regulation with appropriate effector tissues. Although they are diverse in structure, the general mechanism of action of these molecules is similar. Upon delivery to a cell, they interact with specific high-affinity cellular acceptor proteins. Depending on the nature of the ligand, this initial interaction may occur at the cell surface or, alternatively, the ligands may diffuse passively into the cell, where they encounter an intracellular receptor protein. Ultimately, however, these interactions are sufficient to permit nanomolar concentrations of a particular hormone to effect profound alterations in cell phenotype. This review discusses our current understanding of how the ovarian steroids, estrogen and progesterone, transduce their endocrine stimuli to specific genetic targets within responsive cells. In addition, it considers the mechanisms by which antihormones, synthetic compounds that oppose estrogen and progesterone action, manifest their biological activity.

Expression of functional estrogen (ER) and progesterone (PR) receptors within endocrine tissues, such as the uterus, mammary gland, and central nervous system, has been well documented (Clark and Peck, 1979). However, it is now generally accepted that estrogen and progesterone have activities other than those directly related to reproduction. This is particularly evident for estrogen; epidemiological and clinical evidence supports a role for this hormone as an osteoprotective agent in postmenopausal women (Barzel, 1988). Although both ER and PR have been detected in bone cell osteoblasts, osteoclasts, and bone marrow stromal cells, it is likely that ER is the more important transcription factor in bone (Brandi *et al.*, 1993; Chow *et al.*, 1992; Wei *et al.*, 1993). In postmenopausal osteoporosis, it has been hypothesized that withdrawal of estrogen leads to an increase in interleukin-6 production and a concomitant increase in the activity of both osteoclasts and osteoblasts (Jilka *et al.*, 1992; Stein and Yang, 1995). However, this event slightly favors an increase in osteoclast activity leading to a progressive loss of bone (Jilka *et al.*, 1992).

In addition to preservation of bone, estrogen replacement therapy has also been shown to be associated with a sharp reduction in the risk of cardiovascular disease associated with menopause (Eaker *et al.*, 1993). Although the mechanism of action of estrogens on the cardiovascular system has been elusive, it is known that they induce vascular relaxation of coronary and aortic artery preparations (Rosano *et al.*, 1993), lowering coronary artery stenosis and carotid wall thickness. Estrogen's action in this system may be related to alterations in blood lipoproteins, coagulation factors, fibrinolytic parameters, insulin sensitivity factors, or other clinical endpoints that are considered beneficial and appear to protect against atherogenesis, myocardial infarction, and other heart failures (Eaker *et al.*, 1993; Quehenberger *et al.*, 1993; Cid *et al.*, 1994). It is not clear at the molecular level how estrogen manifests these activities. However, the observation that ER is expressed in some myocardial cells and in the smooth muscle cells of arterial walls of the endometrium is consistent with a role for estrogen in these tissues (Stumpf *et al.*, 1977). This unique distribution suggests that ovarian steroids have a direct effect on the cardiovascular system and on vascular blood flow (Perrot-Applanat *et al.*, 1988). The accumulating evidence that ovarian steroid receptors and their ligands are involved in processes other than reproduction has been met with an increased interest in defining the mechanism of action of compounds that modulate the biological activity of these receptors. Emanating from these endeavors are a new generation of steroid receptor modulators, the mechanisms of which are discussed in this review.

II. The Mechanism of Action of Estrogen and Progesterone

A. Estrogen and Progesterone Receptors

The biological activity of estrogen and progesterone is manifested through high-affinity receptors located in the nuclei of target cells. The observation that both ER and PR operate as hormone-dependent transcription factors when expressed in the lower eucaryote *Saccharomyces cerevisae* (McDonnell *et al.*, 1991a; Metzger *et al.*, 1988; Vegeto *et al.*, 1992) demonstrates that the presence of the receptor alone is sufficient to confer upon a cell the ability to respond to hormone administration. Whereas a specific nuclear receptor appears to be required for steroid hormone action in target cells, there is some evidence that the ultimate response of a cell to hormone is manifested by a combination of receptor-mediated genomic and nongenomic activities. For genomic actions, the hormone-activated receptor interacts directly with the regulatory region of a target gene and alters its rate of transcription (Beato *et al.*, 1995). However, in the case of estrogen, the hormone-activated receptor can have additional activities that precede its genomic effects (Aronica *et al.*, 1994; Ince *et al.*, 1994). Specifically, it has been shown that estrogenic compounds can directly stimulate adenyl cyclase activity in ER-containing cells (Aronica *et al.*, 1994). The observation that the steroid specificity of this activity mirrors the known binding affinities of the compounds for ER makes it likely that the receptor is involved in adenyl cyclase activation (Aronica *et al.*, 1994). Although the overall contribution of this novel pathway to ER action remains to be determined, the fact that this process appears to occur also in intact rat uterus provides compelling evidence for its biological significance.

The receptors for estrogen and progesterone are members of a large superfamily of nuclear proteins that mediate the biological action of steroids, thyroid hormone, and vitamins D and A (Mangelsdorf *et al.*, 1995). In addition, this family includes also a large number of receptors for which ligands have not yet been identified (Laudet *et al.*, 1992). Analysis of the sequence of these receptor cDNAs coupled with extensive studies of the biological activity of a great number of receptor mutations has permitted the definition of the functional domains within these proteins (McDonnell et al., 1993). The results and consequences of these studies have been reviewed extensively elsewhere and are only discussed briefly here (Beato et al., 1995; Mangelsdorf et al., 1995). We want to focus specifically on how the availability of these cDNAs and their use as tools to genetically dissect receptor function has helped our understanding of ER and PR pharmacology.

The human estrogen receptor (hER) exists as a single 65-kDa protein within target cells (Greene et al., 1986). Although variant mRNAs that could potentially encode hER variants have been identified, their biological significance has not yet been determined (Fuqua et al., 1991, 1993). Unlike hER, the human progesterone receptor (hPR) occurs as two distinct forms within target cells, hPR-A and hPR-B, of 94 and 114 kDa, respectively (Horwitz and Alexander, 1983). The cloning and characterization of the cDNAs corresponding to these receptors indicated that hPR-B differs from hPR-A only in that it contains an additional amino-terminal fragment of 164 amino acids (B164) (Kastner et al., 1990b). The subsequent use of these cloned cDNAs to establish reconstituted progesterone-dependent transcription systems in heterologous cells has permitted a determination of specific roles for each of these two PR subtypes (discussed below). Both forms of PR are derived from a single gene as a consequence of alternate initiation of transcription from distinct promoters (Kastner et al., 1990). Northern blot analysis revealed the existence of at least six distinct mRNAs, indicating that the transcriptional regulation of the PR-gene promoters *in vivo* is quite complex (Gronemeyer et al., 1991; Wei et al., 1988). The initiation start sites of the most abundant of these mRNAs have been mapped, indicating that multiple messages are derived from each of the two PR promoters, giving rise to several hPR-A- and hPR-B-specific transcripts. The significance of multiple transcripts for each receptor isoform remains to be determined.

Despite the noted differences in PR and ER, they share the same overall structure. Interestingly, although the average molecular masses for ER and PR ligands are approximately 300 Da, respectively, it has been determined that more than 300 amino acids of the carboxyl terminus are required for high-affinity ligand binding (McDonnell et al., 1993). The implied complexity of the hormone-binding region of this family of receptors was confirmed by the completion of the crystallographic structural analysis of the ligand-binding domains of the RXR and thyroid hormone receptors (Renaud et al., 1995; Wagner et al., 1995). Within the ligand-binding domains are subdomains that have been shown to be responsible for interaction with heat shock proteins, dimerization, and nuclear translocation. The role of these domains in receptor action is discussed later.

The smaller DNA-binding domain consists of a 66 to 68-amino acid region that contains nine cysteines that are invariant among the nuclear receptors (Laudet et al., 1992). These cysteins have been shown to lie at the base of two separate zinc finger motifs (Klug and Rhodes, 1987). When the structure of the DNA-binding domain of ER, and the related glucocorticoid receptor (GR)-binding domain, were analyzed crystallographically, it was shown that the DNA-binding competent form of the receptor was a dimer and that in each dimer the first finger of each monomer binds to the DNA target, placing each subunit into adjacent major grooves of the double helix (Freedman et al., 1988; Luisi et al., 1991; Schwabe et al., 1990). Additional studies have shown that the base of this finger, the P-box, endows upon the receptor the ability to recognize different target sequences on promoters (Umesono and Evans, 1989). The second finger appears not to be involved directly in contacting DNA but is involved in the formation of intimate contacts between receptor dimers and possibly with other associated proteins (Luisi et al., 1991).

The amino terminus is the most divergent domain among the receptors and appears to be involved mainly in making contacts with the general transcription machinery (Danielian et al., 1993; Danielsen et al., 1987; Dobson et al., 1989; Hollenberg and Evans, 1988; Kastner et al., 1990; Tasset et al., 1990). Although the precise location of the sequences required for transactivation have not yet been identified, it has been proposed that in ER a single transactivation function is located within this region (Danielian et al., 1993; Tasset et al., 1990; Tzukerman et al., 1994), as is the case for the A isoform of hPR (Kastner et al., 1990), whereas in hPR-B an additional B-upstream sequence seems to be required for maximal transcriptional activity (Meyer et al., 1992; Sartorius et al., 1994).

B. Transcriptional Regulation by Estrogen and Progesterone Receptors

The overall mechanism of action of steroid hormone receptors is similar (Beato et al., 1995; O'Malley et al., 1995). This general mechanism is discussed next, followed by a consideration of the specific aspects of hER and hPR action that may be different. In the absence of hormone, the steroid hormone receptors reside in the nucleus of target cells in a latent form associated with a high-molecular-weight complex comprising heat shock proteins 90 (hsp90), 72 (hsp72), and 59 (hsp59)

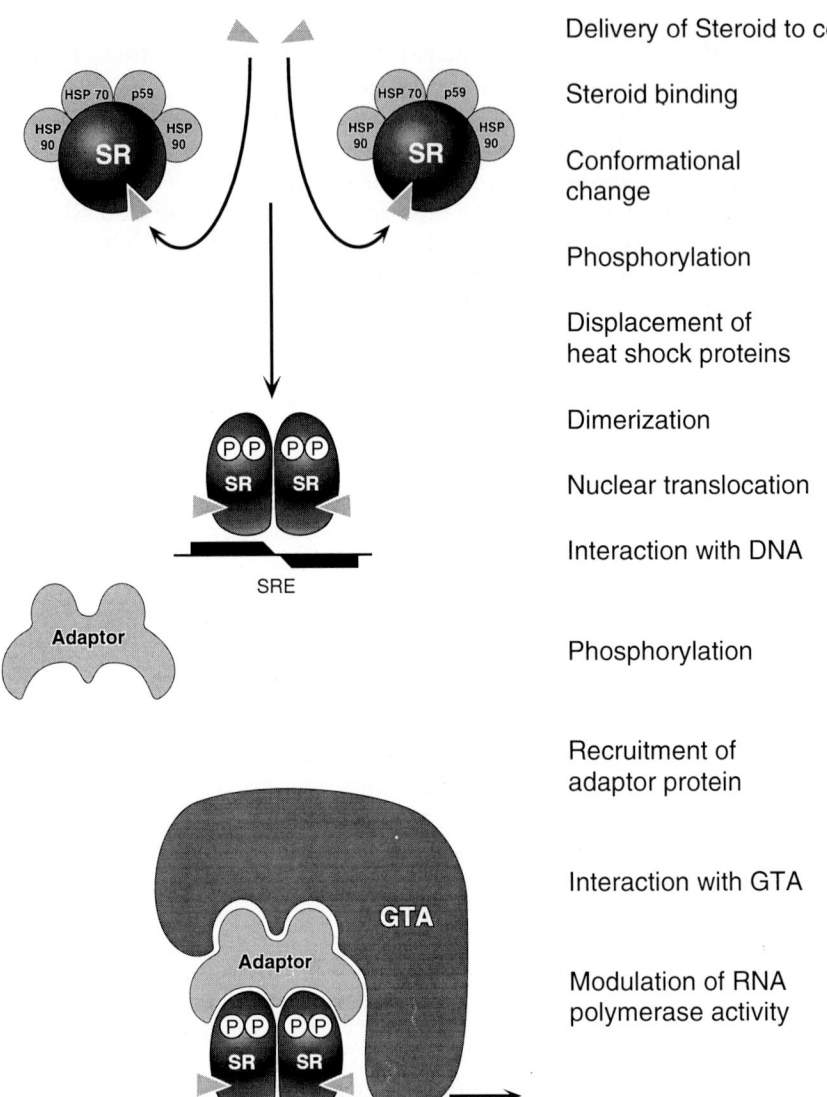

FIG. 1. The mechanism of action of steroid hormones. Steroid hormones exert their effects on gene transcription via specific intracellular receptor proteins. Genetic and biochemical evidence suggests that signal transduction to the nucleus occurs in a series of distinct steps. The details of the model are discussed in the text; in brief, hormone enters the cell passively, where it encounters its cognate receptor (SR) in a complex with heat shock proteins (HSPs). The binding of ligand initiates a cascade of molecular events, including phosphorylation, dimerization, nuclear translocation, interaction with

and possibly other proteins (Bagchi *et al.*, 1991; Pratt, 1990; Smith *et al.*, 1990) (Fig. 1). The precise stoichiometry of the individual components within the receptor–hsp complex is difficult to determine; however, it is generally considered that only a single receptor molecule exists in each oligomeric complex (Renoir *et al.*, 1990). Although the association of steroid receptors with hsp90 *in vitro* has been well established, it is only recently that these proteins have been shown to interact *in vivo*. Using a recombinant chicken hsp90 that was modified to contain a nuclear localization signal, it has been demonstrated that a mutant GR, which lacked a nuclear localization signal, could be translocated to the nucleus, implying that these proteins contact each other within the cell (Kang *et al.*, 1994). One caveat to these experiments was that, as a consequence of the expression system used, the recombinant hsp90 was greatly overexpressed, representing about 20% of the total cellular protein. This raises the possibility that the observed nuclear translocation of the recombinant hsp90 in this system did not reflect what occurs normally within the cell.

The role of hsp in steroid receptor action is unknown; however, they may be involved in (1) assisting the folding of the nascent receptor peptide or (2) maintaining the receptor in a transcriptionally inactive state in the absence of hormone (Cadepond *et al.*, 1991; Kost *et al.*, 1989; Pratt *et al.*, 1992). It is currently considered that the reversible binding of the steroid ligand defines a conformational modification of receptor structure that releases the receptor from the hsp complex (Bagchi *et al.*, 1991). The role of hsp90 as a general repressor of steroid hormone receptor transcriptional activity has been challenged by the observation that hPR free of associated hsp90 still requires hormone for transcriptional activity when assayed in a hormone-dependent transcription system reconstituted *in vitro* (Bagchi *et al.*, 1992). Although all the steroid hormone receptors have been shown to be associated with hsp90, it is likely that its role in the biology of each receptor is different. For instance, it has been shown that hsp90 is required for the formation of a hormone-binding competent form of GR (Bresnick *et al.*, 1989), but it does not appear to be absolutely required for correct folding of ER (Picard *et al.*, 1990).

Very little is known about the role of the other hsp that are associated

---

specific DNA response elements (SREs), and recruitment of adaptor proteins that allow the steroid receptor to productively interact with the general transcription apparatus (GTA). The transcriptional effects of the hormone on RNA polymerase activity are determined ultimately by the cellular and promoter contexts of the receptor bound to DNA.

with the untransformed receptor. The most tightly associated protein, hsp72, does not seem to be released from the receptor upon activation with hormone alone and *in vitro* at least requires additional treatment with ATP (Bagchi *et al.*, 1992). In a reconstituted nuclear transport system, it has been shown that this protein is involved in the translocation of the simian virus-40 T antigen to the nucleus (Shi and Thomas, 1992). It may be involved in steroid hormone action in a similar way. Several excellent reviews have discussed the potential roles of the other hsp associated with the steroid receptors (Pratt, 1990; Smith and Toft, 1993).

C. Initiation of the Progesterone Receptor-Mediated Signal Transduction Pathway

Interaction of progesterone with PR initiates the signal transduction cascade by promoting displacement of hsp (DeMarzo *et al.*, 1991) and facilitating the formation of stable receptor dimers. In cells in which hPR-B and hPR-A are coexpressed, three distinct types of dimer (A:A, A:B, B:B) can form; the relative concentration of each dimer pair formed is directly proportional to the expression level of hPR-A and hPR-B (DeMarzo *et al.*, 1991; Edwards *et al.*, 1989). The possible biological significance of these dimer pairs is discussed in detail later. The activated receptor dimers are then capable of interacting with high affinity with specific progesterone response elements located within target gene promoters (Bagchi *et al.*, 1988). In addition to promoting the formation of receptor dimers, the interaction of receptor with hormone facilitates an increase in the overall phosphorylation state of PR (Bagchi *et al.*, 1992; Takimoto *et al.*, 1992; Takimoto and Horwitz, 1993). This appears to occur in two discrete steps, one phosphorylation event occurring upon displacement of hsp and the second occurring following the association of the receptor with DNA (Takimoto *et al.*, 1992; Takimoto and Horwitz, 1993). As yet, a specific role for phosphorylation is not known.

D. Initiation of the Estrogen Receptor-Mediated Signal Transduction Pathway

The association of ER with specific DNA sequences in the regulatory regions of target genes is a prerequisite step in its signal transduction pathway. However, the role of hormone in promoting these receptor–DNA interactions is controversial (McDonnell *et al.*, 1991b). Analysis of ER–DNA interactions *in vitro* using a band shift assay, for instance, indicates that hormone is not required for specific DNA binding (Dana et al., 1994). Additionally, it has been shown that transfection of an ER-expression vector into mammalian HeLa cells with an ER-responsive reporter construct permits some ligand-independent activation of target gene transcription to occur in the absence of added ligand (Tzukerman et al., 1990). It is possible that this type of ligand-independent activation occurs as a consequence of receptor overexpression. However, it does suggest that in some cell and promoter contexts ER can interact with DNA and regulate gene transcription in a ligand-independent manner. In contrast to these data, several studies from our laboratory and others have indicated that hormone is required to promote a high-affinity association of the receptor with DNA in intact cells (McDonnell et al., 1991b; Pham et al., 1991a,b, 1992).

E. STEROID HORMONE RECEPTORS AS TRANSCRIPTION FACTORS

Significant advances in defining the mechanism of transcriptional regulation by RNA polymerase II have been achieved (Tjian and Maniatis, 1994), providing insights that are likely to enhance our understanding of steroid hormone action. Briefly, eucaryotic transcription relies on a complex network of protein–protein interactions that occur between proteins bound at specific DNA sites within a particular promoter. These regulatory sequences are located near the start site of transcription (core promoter) and at remote positions upstream or downstream thereof (enhancer). The core promoter, which is required for basal transcription, consists of the initiation site at +1 (Inr) and the TATA box around −25 (Smale, 1994). The TATA box is bound by transcription factor IID (TFIID), a large multiprotein complex comprising the TATA box-binding protein (TBP) and several TBP-associated proteins (TAFs). In addition, TFIID can interact with the general transcription factors TFIIA, TFIIB, TFIIF, TFIIH, and RNA polymerase II and is responsible for their recruitment to the start site of transcription, where the preinitiation complex is formed (Conaway and Conaway, 1993; Goodrich and Tjian, 1994). In the subsequent stages of basal transcription, the preinitiation complex is disrupted and the polymerase is released for elongation of the transcript. TFIID constitutes a docking complex for the recycling of the other general transcription factors, permitting further rounds of transcription (Zawel et al., 1995).

In activated transcription, the formation of a preinitiation complex is facilitated by gene-specific transcription factors (or transactivators) such as nuclear receptors, which bind cognate regulatory DNA sequences at positions in the proximal promoter or at the remote enhancers (Mitchell and Tjian, 1989). Specific transcription factors are typically composed of a DNA-binding domain and a transcription activation domain. Depending on the nature of this activation domain, specific protein–protein contacts can occur between the transactivator and specific targets such as a TAF protein within TFIID (Chen et al., 1994) or with other protein components of the general transcription machinery (Roberts and Green, 1994). This process is thought to stabilize the preinitiation complex and to increase the rate of reinitiation. In addition to direct contacts between enhancer-binding proteins and the general transcription machinery, it has been shown that indirect contacts via coactivators or adaptors (Lewin, 1990; Ptashne, 1988; Ptashne and Gann, 1990) may also be important on some promoters. The most compelling evidence for these coactivators was collected from experiments in which overexpression of one transcription factor leads to diminished activity of another transactivator with a related activation domain (Martin et al., 1990). This process, referred to as "squelching" or "transcriptional interference," implies that the availability of such coactivators is limited and that regulation of their production can impact on the transactivation potential of an upstream activator. Proteins with coactivator properties have been identified from yeast (Piña et al., 1993; Struhl, 1993), viruses (Desaintes et al., 1992; Haviv et al., 1995), and mammalian cells (Chrivia et al., 1993; Ge and Roeder, 1994; Kretzschmar et al., 1994; Le Douarin et al., 1995; Wilson et al., 1993; Yu et al., 1995). In addition, tissue-specific expression of some coactivators can endow cell specificity to ubiquitously expressed transactivators, as in the case of OBT-1/Bob-1 and Oct-1 (Gstaiger et al., 1995; Strubin et al., 1995).

The precise mechanism by which steroid receptors alter transcription by RNA polymerase II is unknown. However, it has been shown that transcriptionally active PR can potentiate target gene transcription, in a reconstituted system in vitro, by stabilizing the preinitiation complex (Klein-Hitpass et al., 1990). Additionally, the demonstration that hER and the B isoform of hPR can contact the general transcription factor TFIIB directly (Ing et al., 1992) indicates that at least part of the receptors' ability to regulate target gene transcription is as a result of contact between the receptor dimer and proteins of the general transcription apparatus. Our understanding of how these interactions affect gene transcription comes largely from investigations of the mechanisms of action of basal and activated transcription that have been done in other fields.

Transcriptional interference experiments suggest that, in addition to general transcription factors, specific coactivators may be required for receptor transcriptional activity in some cell and promoter contexts (Hoeck et al., 1992; Meyer et al., 1992; Tasset et al., 1990). It has been postulated that these factors may impart upon the transcription machinery the ability to distinguish between agonist- and antagonist-activated receptor and, as a consequence, convey a molecular interpretation of a hormonal stimulus to the transcription apparatus in a promoter- and cell-specific manner (McDonnell et al., 1995; Tzukerman et al., 1994). Candidate adaptor proteins that interact specifically with agonist- but not antagonist-activated hER have been identified biochemically (Cavaillès et al., 1994, 1995; Halachmi et al., 1994; Jacq et al., 1994; McDonnell et al., 1995; Tzukerman et al., 1994). The cDNA for one of these proteins, RIP140, has been isolated (Cavaillès et al., 1995). It has been proposed that this protein is a component of the machinery that permits a cell to distinguish between estrogen, an agonist, and tamoxifen, a cell-specific receptor antagonist. However, no functional data to support this contention are yet available. Additional proteins, such as the mouse protein TIF1, have been shown to interact with hormone-activated ER and related nuclear receptors in vitro (Le Douarin et al., 1995). However, the physiological role of these in vitro biochemical interactions in nuclear receptor action remains to be determined.

III. STEROID HORMONE RECEPTOR ANTAGONISTS

A. INSIGHTS INTO ANTIHORMONE ACTION FROM OTHER SYSTEMS

It is interesting how nature has avoided creating antihormones, like hormones, that can interact with a given receptor and manifest activity opposite to that of its hormones. It appears that nature instead has derived additional mechanisms to regulate the biological activity of hormones. These include altering hormone biosynthesis, metabolism, and secretion; modulating uptake and reuptake (as in the case of neurotransmitters); or downregulating the acceptor proteins that mediate the activities of the hormone. Thus, antihormones, as we know them today, are synthetic compounds that have been developed to oppose the natural action of a given hormone. Considering all receptor-based systems for which antagonists have been developed, three basic modes of action are apparent: modulation of (1) signal surge; (2) signal duration, by altering hormone synthesis and metabolism; and (3) receptor activity, by competing with the endogenous substance for binding to the signal-transducer protein (the receptor) or altering the expression level of the receptor. Steroid receptor antihormones thus represent one mechanism that can be used to regulate the action of endogenous hormones *in vivo*.

The antihormones reviewed in this article interact directly with high affinity with either ER or PR and inhibit the biological activity of estrogens and progestins, respectively, when assayed *in vivo*. Because of the widespread use and success of antiestrogens and antiprogestins in the clinical setting, there is now substantial interest in defining their molecular mechanism of action. This was further fueled by the cloning of ER and PR and their use to create hormone-responsive transcription systems *in vitro* and *in vivo* (Le Douarin *et al.*, 1995; McDonnell *et al.*, 1993). Cumulatively, these studies in the clinic and in more basic aspects of steroid receptor action have significantly enhanced our understanding of antihormone action (McDonnell, 1995). Undoubtedly, these insights will have a major impact on the development of new clinically important receptor modulators.

B. THE MOLECULAR MECHANISM OF ACTION OF ANTIHORMONES

Conceptually, antihormones could be developed that function in a competitive manner to block the access of the endogenous ligands to their receptor. Consequently, the steroid hormone receptor would remain in an unactivated, latent state within the cell (Fig. 2A) (Clark and Peck, 1979). Alternatively, antihormones could function as "pseudoagonists" by mimicking some of the actions of agonists but ultimately pushing the steroid receptor down a transcriptionally nonproductive pathway. The product of this reaction, the "inactive" receptor (Fig. 2B), may have additional activities by competing with the agonist-activated steroid receptor for DNA-binding sites and/or for components of the transcriptional machinery. The pharmacological implications of these two types of antagonists are likely to be different (McDonnell, 1995). Given our current understanding of antihormone action, the second model likely explains the mechanism of action of the known steroid receptor antagonists.

Although agonists and antagonists appear to be equally effective in activating steroid hormone receptors and promoting dissociation of hsp (Bagchi *et al.*, 1988), it is clear that they have unique effects on receptor structure and function (McDonnell *et al.*, 1995; Tzukerman *et al.*, 1994). For some time it has been postulated that interaction with agonists or antagonists induces different conformational changes within the occupied receptor. Analysis by sucrose density gradient centrifugation (De Marzo *et al.*, 1991; Beck *et al.*, 1993; El-Asry *et al.*, 1989), DNA

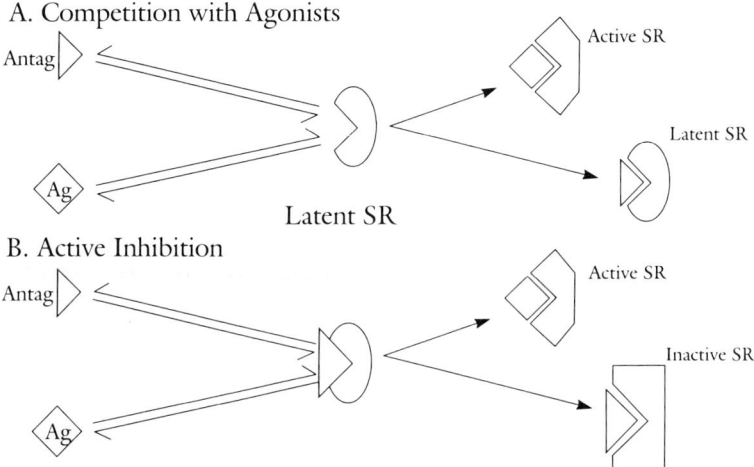

FIG. 2. Potential mechanisms of action of steroid hormone receptor antagonists. In the absence of hormone, the steroid receptor (SR) resides in an inactive complex in the nuclei of target cells. Hormone binding to the receptor initiates a cascade of events that result ultimately in the alteration of the rate of transcription of a target gene. Given what we know of the signal transduction pathway of sex steroids, it is likely that at least two types of antagonists (Antag) can be developed. One class of antagonists could competitively interact with the receptor and block access to agonists (Ag). This passive type of inhibition would maintain the receptor in a "latent state" while the antagonist is physically bound to the receptor. Alternatively, inhibition could be more "active." In this mode antagonists would behave as pseudoagonists, mimicking some of the effects of agonists. Thus, the antagonist would convert the SR from a latent form to one that is "inactive." This form of the receptor is likely to have additional inhibitory activities by competing with hormone-activated SR for DNA-binding sites and ultimately by associating nonproductively with the general transcription apparatus. The currently available antihormones all appear to function as active inhibitors or pseudoagonists.

band shift analysis (El-Ashry et al., 1989), conformation-specific antibodies (Weigel et al., 1992), or differential sensitivity to chemical modifying agents (Moudgil et al., 1989) supports this hypothesis. Direct evidence that the PR and ER ligand-binding domains assume different conformations when bound to agonists or antagonists was provided by the observation that these receptor–ligand complexes are differentially sensitive to the action of proteases (Allan et al., 1992; Beekman et al., 1993; McDonnell et al., 1995; Vegeto et al., 1992). Additional studies that have examined intracellular trafficking of ligand-activated receptors (Dauvois et al., 1992; Dauvois and Parker, 1993), receptor dimerization (Guichon-Mantel et al., 1989), and receptor-induced alterations in chromatin structure (Pham *et al.*, 1991b) demonstrate that these biochemical differences have a functional consequence within the cell.

Important hints as to how antiprogestins operate were provided by Guichon-Mantel *et al.* (1988), who demonstrated that RU486 (an antiprogestin) could promote high-affinity interactions of PR with DNA at target genes. Additional studies have confirmed that antihormone-activated receptors can interact with DNA (Bagchi *et al.*, 1990; Meyer *et al.*, 1990). Therefore, delivery of the receptor to DNA and the biochemical reactions that occur following displacement of hsp are not sufficient to induce receptor transcriptional activity. Thus, it appears that it is the cellular processes that operate downstream of DNA binding that have the ability to distinguish agonist- from antagonist-activated receptors (McDonnell and Goldman, 1994; McDonnell *et al.*, 1992, 1994a, 1995; Tzukerman *et al.*, 1994). Although the general mechanisms of antagonism of the steroid hormone receptors by antihormone are similar, there are some important differences. These are discussed in the following sections.

C. Antagonism of Progesterone Receptor Function

All of the currently available antiprogestins interact directly with the hormone-binding domain of PR and competitively inhibit progesterone binding. This event alone is insufficient, however, to completely block receptor activation as both agonists and antagonists of PR promote displacement of hsp, permit dimerization, and facilitate association of the receptor with DNA (Takimoto *et al.*, 1992). Thus, it appears that antiprogestins must block PR-mediated transcriptional activity at some step downstream of DNA binding. All but one antiprogestin, ZK98299 (onapristone), appear to function in this way (Bocquel *et al.*, 1993; Klein-Hitpass *et al.*, 1993). The mechanism of action of ZK98299 is distinct in that it binds to receptor but does not promote the formation of a high-affinity PR–DNA complex when assayed *in vitro*. One possible interpretation of this result is that ZK98299 prevents receptor dimerization, a requisite step for DNA binding. Although there are as many reports supporting this contention as opposing it, there has been sufficient controversy to prompt the classification of antihormones as type I, which prevent DNA binding, and type II, which appear to deliver the receptor to DNA (Table I). In addition, unlike onapristone (type I), the type II antiprogestins appear to exhibit partial agonist activities under some experimental conditions. Recently, however, the type I antiprogestin ZK98299 has been

TABLE I
Antiprogestin Types Revealed by *in Vitro* Analysis

Type	Function	Structure
I	Prevent or impair receptor–DNA interaction	ZK98299 (onapristone)
II	Interfere with a required process following DNA binding	RU486 (mifepristone)

found to have a lower affinity than most of the other type II antiprogestins, suggesting possibly that the distinction between type I and type II antiprogestins, is not a mechanistic distinction but merely reflects a difference in affinity. This possibility is in agreement with the data from Milgrom's laboratory, which demonstrate that *in vivo,* at concentrations of ligand that saturate the receptor, ZK98299 is functionally identical to other antiprogestins (Delabre *et al.,* 1993). Irrespective of these results, however, the current basis for classifying PR antagonists is whether they prevent (type I) or promote (type II) the association of receptor with DNA *in vitro* (using the nomenclature of Klein-Hitpass *et al.,* 1993).

In support of the *in vitro* classification of antiprogestins as either type I or type II, several groups have shown that in some cell and promoter contexts the pharmacology of PR antagonists can be altered by addition of cAMP analogues (Beck *et al.*, 1993; Sartorius *et al.*, 1993). In particular, it has been shown that type II antiprogestins (but not type I) can function as PR agonists in the presence of 8-Br-cAMP (a cAMP analogue) (Beck *et al.*, 1993). Additionally, it has been shown that DNA-binding activity is required for these responses to cAMP. Thus, it appears that antagonist-activated PR, associated with its DNA response element, is a target for cAMP-stimulated processes that act on the receptor, allowing it to activate transcription. It will be interesting to determine whether the mechanistic differences displayed by antiprogestins *in vitro* are reflected by distinct biological activities *in vivo* (Klein-Hitpass *et al.*, 1993).

An important clue to understanding how the cellular transcriptional machinery distinguishes between PR agonists and antagonists was provided by the elegant studies of Allan *et al.* (1992). By performing limited protease digestion of *in vitro*-synthesized PR in the absence or presence of ligands, it was demonstrated that progesterone and RU486 induce distinct conformational changes within the receptor protein. Using specific monoclonal antibodies, this conformational change was shown to occur at the extreme carboxyl terminus of the receptor (Allan *et al.*, 1992; McDonnell *et al.*, 1994a; Vegeto *et al.*, 1992). Thus, the ability of the transcriptional machinery to distinguish between these agonist- and antagonist-induced structures may be a critical determinant of the biological activity of these compounds. Analysis of receptor structure by protease digestion has been performed on a great number of compounds in different laboratories (Allan *et al.*, 1992; Clemm *et al.*, 1995; McDonnell *et al.*, 1994c). Interestingly, regardless of their chemical derivation, all agonists and antagonists afford protection from trypsin treatment of either a 30- or a 27-kDa receptor fragment, respectively (Clemm *et al.*, 1995). These data firmly support the original hypothesis set forth by Allan *et al.* (1992) that receptor agonists and antagonists induce distinct structural alterations within PR, and suggest further that it is the ability of the cellular transcriptional machinery to recognize these distinct receptor conformations that determines agonist and antagonist efficacy. One caveat to this conclusion, however, is that the chemical derivation of the progestins and antiprogestins available is very similar, and predictions about mechanism may not extend to antiprogestins of different chemical classes.

Using a progesterone-responsive transcription unit in yeast in which intact hPR-B is expressed, the structural elements within PR that discriminate between agonist and antagonist function have been determined. Specifically, using this approach, a mutant PR (PR-UP-1), in which RU486 but not progesterone functioned as a receptor agonist, was identified. It was determined that this phenotype resulted from a truncation of 42 amino acids from the carboxyl terminus of PR (Vegeto et al., 1992). Hormone-binding analysis indicated that RU486 but not progesterone could interact with the mutant receptor. This information suggested that progesterone and RU486 do not interact with PR in the same manner. When assayed in transiently transfected mammalian cells, the transcriptional activity of the PR-UP-1 mutant receptor was stimulated by RU486 but not progesterone, as was observed in yeast (Vegeto et al., 1992). This important information indicated that the protein sequences required for progesterone and RU486 binding were distinct, and that the carboxyl tail of the receptor may be part of a functional domain of PR responsible for maintaining the receptor in a transcriptionally inactive form in the absence of hormone. Thus it appears that it is the ability of the cellular transcription apparatus to distinguish between the agonist- and antagonist-induced alterations in the structure of the carboxyl tail that determine the biology of these ligands.

Using the information gained from the biochemical and genetic experiments detailed earlier, we have developed a working model to explain the mechanism by which PR distinguishes between agonists and antagonists (Fig. 3). In this model we propose that the carboxyl-terminal region of PR acts as a transcriptional repressor. This event results from either inter- or intramolecular interactions. We propose that, in the presence of PR agonists, a conformational change occurs within the receptor that disrupts the inhibitory effects of the tail region, thus facilitating interactions critical for transactivation. It is further considered that the conformational changes occurring within PR following antagonist binding permit displacement of hsp but are not sufficient to overcome the effect of the inhibitory tail domain. However, since it appears that the UP-1 mutated receptor remains associated with hsp in the absence of hormone (E. Vegeto and D. P. McDonnell, unpublished results), the transcriptional activity of this mutant is manifested only when the protein is delivered to DNA. Since the agonists tested do not interact with PR-UP-1, they are unable to displace the hsp. However, antagonists such as RU486 are capable of performing this task and so appear to be transcriptional activators. It is therefore likely that any compounds (agonists or antagonists) that interact with the PR-UP-1 protein and facilitate hsp displacement will function as agonists. Interestingly, a model similar to ours has evolved from studies of the mechanism of action of the VP16 acidic activator (Roberts and Green, 1994). Specifically, it was demonstrated that the activity of TFIIB is inhibited by an intramolecular inhibition involving

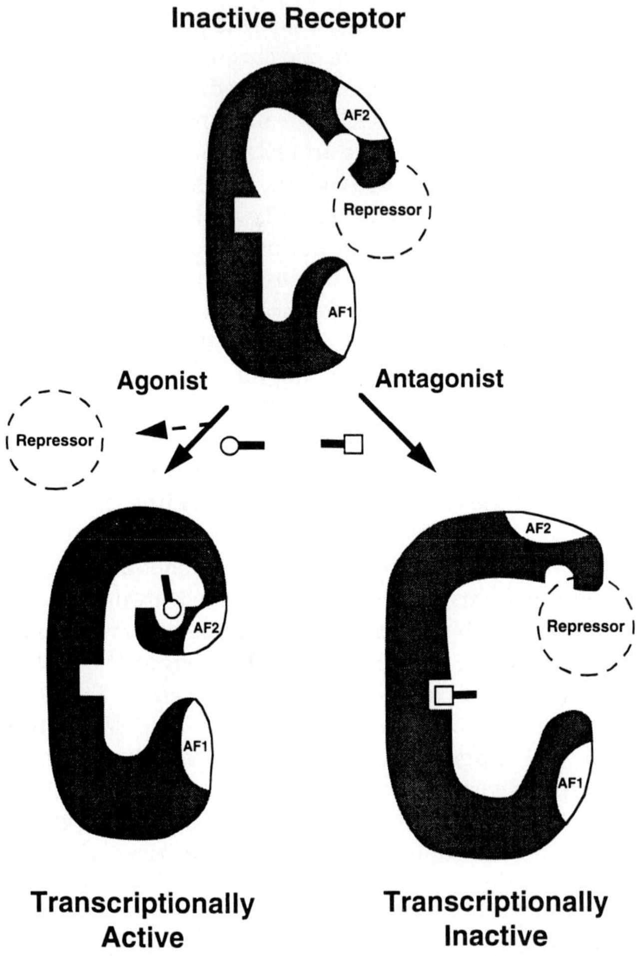

Inactive Receptor

Agonist   **Antagonist**

Transcriptionally Active   **Transcriptionally Inactive**

FIG. 3. The mechanism of action of PR agonists and antagonists. Based on information that has been published (and referenced in the text) and additional data presented in this review, we propose the model illustrated here to explain how the cell distinguishes between progesterone receptor agonists and antagonists. In the absence of ligand, PR resides in the nucleus in a transcriptionally inactive state, associated with a repressor protein. This repressor protein inhibits PR transcriptional activity by interacting with the carboxyl-tail of the receptor, a region that appears to be required for function. Upon interaction with an agonist, the receptor undergoes a conformational change facilitating the displacement of this repressor protein and permitting PR to manifest transcriptional activity. The interaction of an antagonist with PR also induces a conformational change within PR; however, this change is not identical to that induced by agonists and appears to be insufficient to promote repressor displacement.

sequences in both ends of the protein. The VP16 acidic activator is capable of disrupting this interaction by inducing a distinct conformational change within TFIIB, permitting its interaction with TFIIF and RNA polymerase. Whether or not the carboxyl tail of PR functions analogously remains to be determined.

D. ANTAGONISM OF ESTROGEN RECEPTOR FUNCTION

Several classes of chemically distinct antiestrogens have been developed. Some of the most commonly used antiestrogens in both therapy and experimental endocrinology are shown in Table II. This group of compounds consists of both steroidal (ICI 164,384) and nonsteroidal (tamoxifen) antiestrogens (Fawell et al., 1990; McDonnell et al., 1995; Tzukerman et al., 1994; Wakeling et al., 1991). The pure antiestrogen, ICI 164,384 is steroidal in nature and was originally developed as an affinity ligand for ER. It was subsequently determined that the addition of the large 7α-alkyl side chain to the steroid nucleus permitted its interaction with the receptor and its subsequent activity as an antiestrogen. Specifically, the amide function and the distance between the amide group and the core steroid nucleus structure have been observed to be important determinants of its antiestrogenic action. The biology of this class of antiestrogens has been studied extensively as they represent the first compounds developed that are devoid of ER partial agonist activity. A second series of pure antiestrogens, the 2-phenylindoles, have been recently reported to be pure antiestrogens in cell culture and *in vivo* (vonAngerer et al., 1994). However, little is known about their mechanism of action. The other widely used antiestrogens are the triphenylethylenes, such as tamoxifen or droloxifene, and the benzothiophene-derived compound raloxifene. These compounds are unusual in that they function as ER antagonists in most tissues, whereas in some cell and promoter contexts they manifest partial agonist activity. In particular, these compounds exhibit antiestrogenic activities in the breast, whereas they manifest estrogenic actions in bone and in the cardiovasccular system. The molecular basis for this action has been studied in extensive detail and is not reviewed further here (Berry et al., 1990; Dana et al., 1994; Ing et al., 1992; McDonnell et al., 1994a,b, 1995; Tzukerman et al., 1994).

The mechanism of action of the different antiestrogens is slowly being unraveled. The steroidal antiestrogen ICI 164,384 is a pure antiestrogen both *in vivo* and *in vitro*. It has been postulated that ICI 164,384 blocks ER function by impairing receptor dimerization (Dana

TABLE II
Antiestrogen Types Revealed by *in Vitro* Analysis

Type	Agonist–antagonist activity	Structure
II	No partial agonist ativity	**ICI 164,384** X= -(CH$_2$)$_{10}$-C(=O)-N(CH$_3$)(CH$_2$CH$_2$CH$_2$CH$_3$) **ICI 182,780** X= -(CH$_2$)$_9$-S(=O)-CH$_2$CH$_2$CH$_2$-C(F)(F)-C(F)(F)-F
III	Antagonist in most tissues; agonist in bone	**Raloxifene**
IV	Agonist activity in uterus and bone	**4-OH Tamoxifen** X= -C$_6$H$_4$-O-CH$_2$CH$_2$-N(CH$_3$)(CH$_3$)

et al., 1994; Dauvois et al., 1992; Fawell et al., 1990). However, whether or not effective binding of ER to DNA after exposure to ICI 164,384 is possible has been controversial. It has been shown that the stability of

ER dimers may vary depending on the cellular source of ER (Arbuckle et al., 1992) and the experimental conditions used to test receptor dimer–DNA interaction. These differences maay explain why some investigators have been unable to show any effects of ICI 164,384 on ER–DNA interactions (Martinez and Wahli, 1989) but do not explain why in several in vivo systems DNA binding is accomplished by these compounds. Unlike ICI 164,384, the benzothiophene (raloxifene)- and triphenylethylene (tamoxifen)-derived antiestrogens clearly deliver ER to DNA. In most circumstances this activity can lead to antagonism of ER activity (Berry et al., 1990; McDonnell et al., 1995; Tzukerman et al., 1994). However, the agonist–antagonist activity of these latter compounds is influenced by cell and promoter context.

Using a series of novel in vitro models, molecular criteria that clearly distinguish ER agonists from partial agonists and additionally classify the known ER antagonists into three functionally distinct categories have been derived. The behavior of known agonists and antagonists suggests that these classifications are related to distinct ligand-induced structural alterations within ER. A model outlining these classifications is shown in Fig. 4. It is proposed that ER exists in an equilibrium between an inactive and an active state, such that in the absence of ligand the inactive conformation is preferred. Interaction of ER with 17β-estradiol stabilizes the complex in a conformation that facilitates transactivation. The relative agonist–antagonist balance of other ER modulators is determined by the intermediate conformation promoted by the particular compound. Adopting the convention established by Klein-Hitpass et al. (1993), which was originally used to classify antiprogestins, it is now proposed that compounds that prevent ER–DNA interactions be called type I antiestrogens. As yet, unlike the case of PR, a type I antiestrogen has not yet been defined. Based on this nomenclature, it is proposed that ICI 164,384 is a type II antiestrogen that induces a conformation closest to that of the inactive receptor. The ER antagonist keoxifene, which can function as a partial agonist under restricted conditions, is a representative member of type III antiestrogens. Finally, 4-OH-tamoxifen (and other triphenylethylene-derived antiestrogens) represent type IV antiestrogens, which stabilize ER in a conformation that allows it to exhibit transcriptional activity on a limited subset of ER-responsive genes (McDonnell et al., 1995; Tzukerman et al., 1994). Currently, this model demonstrates that the known antiestrogens can be divided into three distinct classes. As new chemicals are identified, it is possible that additional classes of antagonist may be discovered.

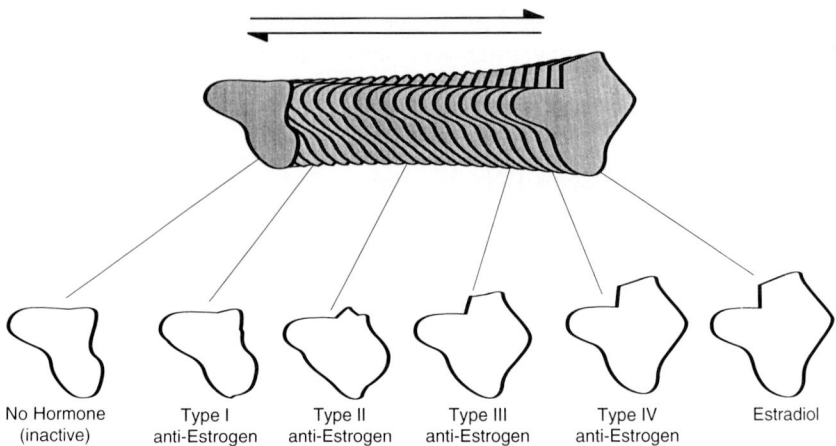

FIG. 4. Hormones and antihormones stabilize distinct conformational states within ER. It is likely that ER can exist in the cell in multiple conformations that represent the inactive state, the active state, and several intermediate states, and that ligands exert their biological activity by stabilizing a specific structure. These distinct conformations could result as a consequence of the ability of compounds to freeze the receptor in a specific conformation by blocking a processive interaction from inactive to active (as shown). Alternatively, the compounds can promote the establishment of unique receptor conformations that result as a consequence of specific ligand–receptor interactions. Based on the nomenclature established by Klein-Hitpass (1993), and the findings of our study, we have divided the known antiestrogens into distinct classes represented by ICI 164,384 (type II), keoxifene (type III), and 4-hydorxytamoxifen (type IV). Although we have been unable to demonstrate the existence of a type I antiestrogen (a compound that prevents receptor–DNA association), conceptually it is likely that such a compound will be discovered. The intense efforts to identify novel ER modulators make it likely that additional types of antiestrogen may emerge in the future.

IV. Final Comments

The information obtained from these and related studies will impact our understanding of the cellular mechanisms that distinguish between hormone agonists and antagonists. In addition, it provides a series of molecular tools with which to predict the *in vivo* biological activity of novel ER and PR modulators. The current state of the art reveals a firm understanding of how agonists and antagonists affect ER and PR structure. The remaining frontier is to define the mechanism by which the cellular transcription machinery distinguishes agonist- from antagonist-activated progesterone receptors. The genetic tools currently available and the ability to reconstitute ER and PR activity *in vitro* will assist greatly in the resolution of this issue.

REFERENCES

Allan, G. F., Leng, X., Tsai, S.-T., Weigel, N. L., Edwards, D. P., Tsai, M.-J., and O'Malley, B. W. (1992). Hormone and antihormone induce distinct conformational changes which are central to steroid receptor activation. *J. Biol. Chem.* **267**, 19513–19520.

Arbuckle, N., Dauvois, S., and Parker, M. (1992). Effects of antioestrogens on the DNA binding activity of oestrogen receptors in vitro. *Nucleic Acids Res.* **20**, 3839–3844.

Aronica, S. M., Kraus, W. L., and Katzenellenbogen, B. S. (1994). Estrogen action via the cAMP signaling pathway: Stimulation of adenylate cyclase and cAMP-regulated gene transcription. *Proc. Natl. Acad. Sci. U.S.A.* **91**, 8517–8521.

Bagchi, M. K., Elliston, J. F., Tsai, S. Y., Edwards, D. P., Tsai, M.-J., and O'Malley, B. W. (1988). Steroid hormone-dependent interaction of human progesterone receptor with its target enhancer element. *Mol. Endocrinol.* **2**, 1221–1229.

Bagchi, M., Tsai, S., Weigel, N., Tsai, M.-J., and O'Malley, B. (1990). Regulation of in vitro transcription by progesterone receptor. *J. Biol. Chem.* **265**, 5129–5134.

Bagchi, M. K., Tsai, S.-Y., Tsai, M.-J., and O'Malley, B. W. (1991). Progesterone enhances target gene transcription by receptor free of heat shock protein hsp90, hsp56, and hsp70. *Mol. Cell. Biol.* **11**, 4998–5004.

Bagchi, M., Tsai, M.-J., O'Malley, B., and Tsai, S. (1992). Analysis of the mechanism of steroid hormone receptor-dependent gene activation in cell-free systems. *Endoc. Rev.* **13**, 525–535.

Barzel, U. S. (1988). Estrogens in the prevention and treatment of postmenopausal osteoporosis. *Am. J. Med.* **85**, 847–850.

Beato, M., Herrlich, P., and Schutz, G. (1995). Steroid hormone receptors: Many actors in search of a plot. *Cell* **83**, 851–857.

Beck, C. A., Weigel, N. L., Moyer, M. L., Nordeen, S. K., and Edwards, D. P. (1993). The progesterone antagonist RU486 acquires agonist activity upon stimulation of cAMP signaling pathways. *Proc. Natl. Acad. Sci. U.S.A.* **90**, 4441–4445.

Beekman, J. M., Allan, G. F., Tsai, S. Y., Tsai, M.-J., and O'Malley, B. W. (1993). Transcriptional activation by the estrogen receptor requires a conformational change in the ligand binding domain. *Mol. Endocrinol.* **7**, 1266–1274.

Berry, M., Metzger, D., and Chambon, P. (1990). Role of the two activating domains of the oestrogen receptor in the cell-type and promoter-context dependent agonistic activity of the anti-oestrogen 4-hydroxytamoxifen. *EMBO J.* **9**, 2811–2818.

Bocquel, M.-T., Ji, J., Ylikomi, T., Benhamou, B., Vergezac, A., Chambon, P., and Gronemeyer, H. (1993). Type II antagonists impair the DNA binding of steroid hormone receptors without affecting dimerization. *J. Steroid Biochem. Mol. Biol.* **45**, 205–215.

Brandi, M. L., Crescioli, C., Tanini, A., Frediani, U., Agnusdei, D., and Gennari, C. (1993). Bone endothelial cells as estrogen targets *Calcif. Tissues Int.* **53**, 312–317.

Bresnick, E. H., Dalman, F. C., Sanchez, E. R., and Pratt, W. B. (1989). Evidence that the 90-kDa heat shock protein is necessary for the steroid binding conformation of the L cell glucocorticoid receptor. *J. Biol. Chem.* **264**, 4992–4997.

Cadepond, F., Schweizergroyer, G., Segardmaurel, I., Jibard, N., Hollenberg, S. M., Giguere, V., Evans, R. M., and Baulieu, E. E. (1991). Heat shock protein-90 as a critical factor in maintaining glucocorticosteroid receptor in a nonfunctional state. *J. Biol. Chem.* **266**, 5834–5841.

Cavaillès, V., Dauvois, S., Danielian, P. S., and Parker, M. G. (1994). Interaction of proteins with transcriptionally active estrogen receptors. *Proc. Natl. Acad. Sci. U.S.A.* **91**, 10009–10013.

Cavaillès, V., Dauvois, S., L'Horset, F., Lopez, G., Hoare, S., Kushner, P. J., and Parker,

M. G. (1995). Nuclear factor RIP140 modulates transcriptional activation by the estrogen receptor. *EMBO J.* **14,** 3741–3751.

Chen, J.-L., Attardi, L. D., Verrijzer, C. P., Yokomori, K., and Tjian, R. (1994). Assembly of recombinant TFIID reveals differential coactivator requirements for distinct transcriptional activators. *Cell* **79,** 93–105.

Chow, J., Thobias, J. H., Colston, K. W., and Chambers, T. J. (1992). Estrogen maintains trabecular bone volume in rats not only by suppression of bone resorption but also by stimulation of bone formation. *J. Clin. Invest.* **89,** 74–78.

Chrivia, J. C., Kwok, R. P. S., Lamb, N., Hagiwara, M., Montminy, M. R., and Goodman, R. H. (1993). Phosphorylated CREB binds specifically to the nuclear protein CBP. *Nature* **365,** 855–859.

Cid, M. C., Kleinman, H. K., Grant, D. S., Shnaper, H. W., Fauci, A. S., and Hoffman G. S. (1994). Estradiol enhances leucocyte binding to tumor necrosis factor (TNF)-stimulated endothelial cells via an increase in TNF-induced adhesion molecules E-selectin, intracellular adhesion molecule type 1, and vascular cell adhesion molecule type 1. *J. Clin. Invest.* **93,** 17–25.

Clark, J. H., and Peck, E. J. (1979). "Female Sex Steroids: Receptors and Function," Vol. 14. Springer-Verlag, New York.

Clemm, B. L., Macy, B., Santiso-Mere, D., and McDonnell, D. P. (1995). Definition of the critical cellular components which distinguish between hormone and antihormone activated progesterone receptor. *J. Steroid Biochem. Mol. Biol.* **53,** 487–495.

Conaway, R. C., and Conaway, J. W. (1993). General initiation factors for RNA polymerase II. *Annu. Rev. Biochem.* **62,** 161–190.

Dana, S. L., Hoener, P. A., Wheeler, D. L., Lawrence, C. L., and McDonnell, D. P. (1994). Novel estrogen response elements identified by genetic selection in yeast are differentially responsive to estrogens and antiestrogens in mammalian cells. *Mol. Endocrinol.* **8,** 1193–1207.

Danielian, P. S., White, T., Hoare, S. A., Fawell, S. E., Parker, M. G. (1993). Identification of residues in the estrogen receptor that confer differential sensitivity to estrogen and hydroxytamoxifen. *Mol. Endocrinol.* **7,** 232–240.

Danielsen, M., Northrop, J. P., Jonklaas, J., and Ringold, G. M. (1987). Domains of the glucocorticoid receptor involved in specific and nonspecific deoxyribonucleic acid binding, hormone activation and transcriptional enhancement. *Mol. Endocrinol.* **1,** 816–822.

Dauvois, S., and Parker, M. G. (1993). "Mechanism of Action of Hormone Antagonists," pp. 166–185. IRL Press, Oxford, England.

Dauvois, S., Danielian, P. S., White, R., and Parker, M. G. (1992). Antiestrogen ICI 164,384 reduces cellular estrogen receptor content by increasing its turnover. *Proc. Natl. Acad. Sci. U.S.A.* **89,** 4037–4041.

Delabre, K., Guiochon-Mantel, A., and Milgrom, E. (1993). In vivo evidence against the existence of antiprogestins disrupting binding to DNA. *Proc. Natl. Acad. Sci. U.S.A.* **90,** 4421–4425.

DeMarzo, A. M., Beck, C. A., Onate, S. A., and Edwards, D. P. (1991). Dimerization of mammalian progesterone receptors occurs in the absence of DNA and is related to the release of the 90-kDa heat shock protein. *Proc. Natl. Acad. Sci. U.S.A.* **88,** 72–76.

Desaintes, C., Hallez, S., Van Alphen, P., and Burny, A. (1992). Transcriptional activation of several heterologous promoters by the E6 protein of human papillomavirus type 16. *J. Virol.* **66,** 325–333.

Dobson, A. D. W., Conneely, O. M., Beattie, W., Mawell, B. L., Mak, P., Tsai, M.-J.,

Schrader, W. T., and O'Malley, B. W. (1989). Mutational analysis of the chicken progesterone receptor. *J. Biol. Chem.* **264,** 4207–4211.
Eaker, E. D., Chesebro, J. H., Sacks, F. M., Wenger, N. K., Whisnant, J. P., and Winston, M. (1993). Cardiovascular disease in women. *Circulation* **88,** 1999–2009.
Edwards, D., Kühnel, B., Estes, P., and Nordeen, S. (1989). Human progesterone receptor binding to mouse mammary tumor virus deoxyribonucleic acid: Dependence on hormone and nonreceptor nuclear factor(s). *Mol. Endocrinol.* **3,** 381–391.
El-Ashry, D., Oñate, S., Nordeen, S., and Edwards, D. (1989). Human progesterone receptor complexed with the antagonist RU486 binds to hormone response elements in a structuraly altered form. *Mol. Endocrinol.* **3,** 1545–1558.
Fawell, S. E., White, R., Hoare, S., Sydenham, M., Page, M., and Parker, M. G. (1990). Inhibition of estrogen receptor-DNA binding by the "pure" antiestrogen ICI 164,384 appears to be mediated by impaired receptor dimerization. *Proc. Natl. Acad. Sci. U.S.A.* **87,** 6883–6887.
Freedman, L. P., Luisi, B. F., Korszun, Z. R., Basavappa, R., Sigler, P. B., and Yamamoto, K. R. (1988). The function and structure of the metal coordination sites within the glucocorticoid receptor DNA binding domain. *Nature (London)* **334,** 543–546.
Fuqua, S. A. W., Fitzgerald, S. D., Chamness, G. C., Tandon, A. K., McDonnell, D. P., Nawaz, Z., O'Malley, B. W., and McGuire, W. L. (1991). Variant human breast tumor estrogen receptor with constitutive transcriptional activity. *Cancer Res.* **51,** 105–109.
Fuqua, S. A. W., Chamness, G. C., and McGuire, W. L. (1993). Estrogen receptor mutations in breast cancer. *Cell. Biochem.* **51,** 135–139.
Ge, H., and Roeder, R. G. (1994). Purification, cloning, and characterization of a human coactivator, PC4, that mediates transcriptional activation of class II genes. *Cell* **78,** 513–523.
Goodrich, J. A., and Tjian, R. (1994). TBP-TAF complexes: Selectivity factors for eukaryotic transcription. *Curr. Opin. Cell Biol.* **6,** 403–409.
Greene, G. L., Gilna, P., Waterfield, M., Baker, A., Hort, Y., and Shine, J. (1986). Sequence and expression of human estrogen receptor complementary DNA. *Science* **231,** 1150–1154.
Gronemeyer, H., Meyer, M. E., Bocquel, M. T., Kastner, P., Turcotte, B., and Chambon, P. (1991). Progestin receptors: Isoforms and antihormone action. *J. Steroid Biochem. Mol. Biol.* **40,** 271–278.
Gstaiger, M., Knoepfel, L., Georgiev, O., Schaffner, W., and Hovens, C. M. (1995). A B-cell coactivator of octamer-binding transcription factors. *Nature (London)* **373,** 360–362.
Guiochon-Mantel, A., Loosfelt, H., Lescop, P., Sar, S., Atger, M., Perrot-Applanat, M., and Milgrom, E. (1989). Mechanisms of nuclear localization of the progesterone receptor:evidence for interaction between monomers. *Cell* **57,** 1147–1154.
Halachmi, S., Marden, E., Martin, G., MacKay, H., Abbondanza, C., and Brown, M. (1994). Estrogen receptor-associated proteins: Possible mediators of hormone-induced transcription. *Science* **264,** 1455–1458.
Haviv, I., Vaizel, D., and Shaul, Y. (1995). The X protein of hepatitis B virus coactivates potent activation domains. *Mol. Cell. Biol.* **15,** 1079–1085.
Hoeck, W., Hofer, P., and Groner, B. (1992). Overexpression of the glucocorticoid receptor represses transcription from hormone responsive and non-responsive promoters. *J. Steroid Biochem. Mol. Biol.* **41,** 283–289.
Hollenberg, S., and Evans, R. (1988). Multiple and cooperative trans-activation domains of the human glucocorticoid receptor. *Cell* **55,** 899–906.

Horwitz, K. B., and Alexander, P. S. (1983). In situ photolinked nuclear progesterone receptors of human breast cancer cells: Subunit molecular weights after transformation and translocation. *Endocrinology* **113,** 2195–2201.

Ince, B. A., Montano, M. M., and Katzenellenbogen, B. S. (1994). Activation of transcriptionally inactive human estrogen receptors by cyclic adenosine 3′, 5′-monophosphate and ligands including antiestrogens. *Mol. Endocrinol.* **8,** 1397–1406.

Ing, N., Beekman, J., Tsai, M.-J., and O'Malley, B. (1992). Members of the steroid hormone receptor superfamily interact with TFIIB (S300-II). *J. Biol. Chem.* **267,** 17617–17623.

Jacq, X., Brou, C., Lutz, Y., Davidson, I., Chambon, P., and Tora, L. (1994). Human $TAF_{II}30$ is present in a distinct TFIID complex and is required for transcriptional activation by the estrogen receptor. *Cell* **79,** 107–117.

Jilka, R. L., Hangoc, G., Girasole, G., Passeri, G., Williams, D. C., Abrams, J. S., Boyce, B., Broxmeyer, H., and Manolagus, S. C. (1992). Increased osteoclast development after estrogen loss: Mediation by interleukin-6. *Science* **257,** 88–91.

Kang, K. I., Devin, J., Cadepond, F., Jibard, N., Guichon-Mantel, A., Baulieu, E. E., and Catelli, M.-G. (1994). In vivo functional protein-protein interaction: Nuclear targeted hsp90 shifts cytoplasmic steroid receptor mutants into. *Proc. Natl. Acad. Sci. U.S.A.* **91,** 340–344.

Kastner, P., Krust, A., Turcotte, B., Stropp, U., Tora, L., Gronemeyer, H., and Chambon, P. (1990). Two distinct estrogen-regulated promoters generate transcripts encoding the two functionally different human progesterone receptor forms A and B. *EMBO J.* **9,** 1603–1614.

Klein-Hitpass, L., Weigel, N. L., Allan, G. F., Riley, D., Rodriguez, R., Schrader, W. T., Tsai, M.-J., and O'Malley, B. W. (1990). The progesterone receptor stimulates cell-free transcription by enhancing the formation of a stable preinitiation complex. *Cell* **60,** 247–257.

Klein-Hitpass, L., Cato, A., Henderson, D., and Ryffel, G. (1993). Two types of antiprogestins identified by their differential action in transcriptionally active extracts from T47D cells. *Nucleic Acids Res.* **19,** 1227–1234.

Klug, A., and Rhodes, D. (1987). Zinc fingers: A novel protein motif for nucleic acid recognition. *Trends Biochem. Sci.* **12,** 464–469.

Kost, S. L., Smith, D. F., Sullivan, W. P., Welch, W. J., and Toft, D. O. (1989). Binding of heat shock proteins to the avian progesterone receptor. *Mol. Cell. Biol.* **9,** 3829–3838.

Kretzschmar, M., Kaiser, K., Lottspeich, F., and Meisterernst, M. (1994). A novel mediator of class II gene transcription with homology to viral immediate-early transcriptional regulators. *Cell* **78,** 525–534.

Laudet, V., Hänni, C., Coll, J., Catzeflis, F., and Stéhelin, D. (1992). Evolution of the nuclear receptor gene superfamily. *EMBO J.* **11,** 1003–1013.

Le Douarin, B., Zechel, C., Garnier, J.-M., Lutz, Y., Tora, L., Pierrat, B., Heery, D., Gronemeyer, H., Chambon, P., and Losson, R. (1995). The N-terminal part of TIF1, a putative mediator of the ligand-dependent activation function (AF-2) of nuclear receptors, is fused to B-raf in the oncogenic protein T18. *EMBO J.* **14,** 2020–2033.

Lewin, B. (1990). Commitment and activation at Pol II promoters: A tail of protein-protein interactions. *Cell* **61,** 1161–1164.

Luisi, B. F., Xu, W. X., Otwinowski, Z., Freedman, L. P., Yamamoto, K. R., and Sigler, P. B. (1991). Crystallographic analysis of the interaction of the glucocorticoid receptor with DNA. *Nature (London)* **352,** 497–505.

Mangelsdorf, D. J., Thummel, C., Beato, M., Herrlich, P., Schutz, G., Umesono, K., Blumberg, B., Kastner, P., Mark, M., Chambon, P., and Evans, R. M. (1995). The nuclear receptor superfamily: The second decade. *Cell* **83**, 835–839.

Martin, K. J., Lillie, J. W., and Green, M. R. (1990). Evidence for interaction of different eukaryotic transcriptional activators with distinct cellular targets. *Nature (London)* **346**, 147–152.

Martinez, E., and Wahli, W. (1989). Cooperative binding of estrogen receptor to imperfect estrogen-responsive DNA elements correlates with their synergistic hormone-dependent enhancer activity. *EMBO J.* **8**, 3781–3791.

McDonnell, D. P. (1995). Unraveling the human progesterone receptor signal transduction pathway: Insights into antiprogestin action. *Trends Endocrinol. Metab.* **6**, 133–138.

McDonnell, D., and Goldman, M. (1994). RU486 exerts antiestrogenic activities through a novel progesterone receptor A form-mediated mechanism. *J. Biol. Chem.* **269**, 11945–11949.

McDonnell, D. P., Nawaz, Z., Densmore, C., Clark, J., Weigel, N. L., and O'Malley, B. W. (1991a). High level expression of biologically active estrogen receptor in Saccharomyces cerevisiae. *J. Steroid Biochem. Mol. Biol.* **39**, 291–297.

McDonnell, D. P., Nawaz, Z., and O'Malley, B. W. (1991b). In situ distinction between steroid receptor binding and transactivation at a target gene. *Mol. Cell Biol.* **11**, 4350–4355.

McDonnell, D. P., Vegeto, E., and O'Malley, B. W. (1992). Identification of a negative regulatory function for steroid receptors. *Proc. Natl. Acad. Sci. U.S.A.* **89**, 10563–10567.

McDonnell, D., Vegeto, E., and Gleeson, M. A. G. (1993). Nuclear hormone receptors as targets for new drug discovery. *Bio/Technology* **11**, 1256–1261.

McDonnell, D. P., Clemm, D., and Imhof, M. O. (1994a). Definition of the molecular mechanisms which distinguish between hormone and antihormone activated steroid receptors. *Semin. Cancer Biol.* **5**, 327–336.

McDonnell, D. P., Clemm, D. L., and Imhof, M. O. (1994b). Definition of the cellular mechanisms which distinguish between hormone and antihormone activated steroid receptors. *Semin. Cancer Biol.* **5**, 503–513.

McDonnell, D. P., Shahbaz, M. S., Vegeto, E., and Goldman, M. E. (1994c). The human progesterone receptor A-form functions as a transcriptional modulator of mineralocorticoid receptor transcriptional activity. *J. Steroid Biochem. Mol. Biol.* **48**, 425–432.

McDonnell, D. P., Clemm, D. L., Herman, T., Goldman, M. E., and Pike, J. W. (1995). Analysis of estrogen receptor function in vitro reveals three distinct classes of antiestrogens. *Mol. Endocrinol.* **9**, 659–669.

Metzger, D., White, J. H., and Chambon, P. (1988). The human oestrogen receptor functions in yeast. *Nature (London)* **334**, 31–36.

Meyer, M. E., Pornon, A., Ji, J., Bocquel, M. T., Chambon, P., and Gronemeyer, H. (1990). Agonist and antagonist properties of RU486 on the functions of the human progesterone receptor. *EMBO J.* **9**, 3923–3932.

Meyer, M.-E., Quirin-Stricker, C., Lerouge, T., Bocquel, M.-T., and Gronemeyer, H. (1992). A limiting factor mediates the differential activation of promoters by the human progesterone receptor isoforms. *J. Biol. Chem.* **267**, 10882–10887.

Mitchell, P. J., and Tjian, R. (1989). Transcriptional regulation in mammalian cells by sequence-specific DNA binding proteins. *Science* **245** 371–378.

Moudgil, V. K., Anter, M. J., and Hurd, C. (1989). Mammalian progesterone receptor shows differential sensitivity to sulfhydral modifying agents when bound to agonist and antagonist ligands. *J. Biol. Chem.* **264,** 2203–2211.

O'Malley, B. W., Schrader, W. T., Mani, S., Smith, C., Weigel, N. L., Conneely, O. M., and Clark, J. H. (1995). An alternative ligand-independent pathway for activation of steroid receptors. *Recent Prog. Horm. Res.* **50,** 333–347.

Perrot-Applanat, M., Groyer-Picard, M. T., Garcia, E., Lorenzo, E., and Milgrom, E. (1988). Immunocytochemical demonstration of estrogen and progesterone receptors in muscle cells of uterine arteries in rabbits and human. *Endocrinology* **123** 1511–1519.

Pham, T. A., Elliston, J. F., Nawaz, Z., McDonnell, D. P., Tsai, M.-J., and O'Malley, B. W. (1991a). Anti-estrogen can establish non-productive complexes and alter chromatin structure at target enhancers. *Proc. Natl. Acad. Sci. U.S.A.* **88,** 3125–3129.

Pham, T. A., Hwung, Y.-P., McDonnell, D. P., and O'Malley, B. W. (1991b). Transactivation functions facilitate the disruption of chromatin structure by estrogen receptor derivatives in vivo. *J. Biol. Chem.* **266,** 18179–18187.

Pham, T. A., Hwung, Y.-P., Santiso-Mere, D., McDonnell, D. P., and O'Malley, B. W. (1992). Ligand-dependent and independent functions of the transactivation regions of the human estrogen receptor in yeast. *Mol. Endocrinol.* **6,** 1043–1050.

Picard, D., Khursheed, B., Garabedian, M. J., Fortin, M. G., Lindquist, S., and Yamamoto, K. R. (1990). Reduced levels of hsp90 compromise steroid receptor action in vivo. *Nature* **348,** 166–168.

Pinña, B., Berger, S., Marcus, G. A., Silverman, N., Agapite, J., and Guarente, L. (1993). ADA3: A gene, identified by resistance to GAL4-VP16, with properties similar to and different from those of ADA2. *Mol. Cell. Biol.* **13,** 5981–5989.

Pratt, W. B. (1990). Interaction of hsp90 with steroid receptors—organizing some diverse observations and presenting the newest concepts. *Mol. Cell. Endocrinol.* **74,** C69–C76.

Pratt, W., Scherrer, L., Hutchison, K., and Dalman, F. (1992). A model of glucocorticoid receptor unfolding and stabilization by a heat shock protein complex. *J. Steroid Biochem Mol. Biol.* **41,** 223–229.

Ptashne, M. (1988). How eukaryotic transcriptional activators work. *Nature (London)* **335,** 683–689.

Ptashne, M., and Gann, A. A. F. (1990). Activators and targets. *Nature (London)* **346,** 329–331.

Quehenberger, P., Kapiotis, S., Partan, C., Schneider, B., and Wenzel, R. (1993). Studies on oral contraceptive-induced changes in blood coagulation and fibrinolysis and the estrogen effect on endothelial cells. *Ann. Hematol.* **67,** 33–36.

Renaud, J.-P., Rochel, N., Ruff, M., Vivat, V., Chambon, P., Gronemeyer, H., and Moras, D. (1995). Crystal structure of the RAR-g ligand-binding domain bound to all-trans retinoic acid. *Nature (London)* **378,** 681–689.

Renoir, J. P., Radanyi, C., Jung-Testas, I., Faber, L. E., and Baulieu, E. E. (1990). The nonactivated progesterone receptor is a nuclear heteroligomer. *J. Biol. Chem.* **265,** 14402–14406.

Roberts, S. G. E., and Green, M. R. (1994). Activator-induced conformational change in general transcription factor TFIIB. *Nature (London)* **371,** 717–720.

Rosano, G. M. C., Sarrel, P. M., Poole-Wilson, P. A., and Collins, P. (1993). Beneficial effect of estrogen on exercise-induced myocardial ischaemia in women with coronary artery disease. *Lancet* **342,** 133–136.

Sartorius, C. A., Tung, L., Takimoto, G. S., and Horwitz, K. B. (1993). Antagonist-occupied human progesterone receptors bound to DNA are functionally switched to transcriptional agonists by cAMP. *J. Biol. Chem.* **268,** 9262–9266.

Sartorius, C. A., Melville, M. Y., Hovland, A. R., Tung, L., Takimoto, G. S., and Horwitz, K. B. (1994). A third transactivation function (AF3) of human progesterone receptors located in the unique N-terminal segment of the B-isoform. *Mol. Endocrinol.* **8,** 1347–1360.
Schwabe, J. W. R., Neuhaus, D., and Rhodes, D. (1990). Solution structure of the DNA-binding domain of the oestrogen receptor. *Nature (London)* **348,** 458–461.
Shi, Y., and Thomas, J. O. (1992). The transport of proteins into the nucleus requires the 70-kilodalton heat shock protein or its cytosolic cognate. *Mol. Cell. Biol.* **12,** 2186–2192.
Smale, S. T. (1994). Core promoter architecture for eukaryotic protein-coding genes. *In* "Transcription: Mechanisms and Regulation," (R. C. Conaway and J. W. Conaway, eds.), Vol. 3, pp. 63–81. Raven Press, New York.
Smith, D., and Toft, D. (1993). Steroid receptors and their associated proteins. *Mol. Endocrinol.* **7,** 4–11.
Smith, D. F., Faber, L. E., and Toft, D. O. (1990). Purification of unactivated progesterone receptor and identification of novel receptor-associated proteins. *J. Biol. Chem.* **265,** 3996–4003.
Stein, B., and Yang, M. (1995). Repression of the interleukin-6 promoter by estrogen receptor is mediated by NF-kB and C/EBPb. *Mol. Cell. Biol.* **15,** 4971–4979.
Strubin, M., Newell, J. W., and Matthias, P. (1995). OBF-1, a novel B cell-specific coactivator that stimulates immunoglobulin promoter activity through association with octamer-binding proteins. *Cell* **80,** 497–506.
Struhl, K. (1993). Yeast transcription factors. *Curr. Opin. Cell Biol.* **5,** 513–520.
Stumpf, W. E., Sar, S., and Aumeller, G. (1977). The heart: A target organ for estradiol. *Science* **196,** 319–321.
Takimoto, G. S., and Horwitz, K. B. (1993). Progesterone receptor phosphorylation: Complexities in defining a functiona role. *Trends Endocrinol. Metab.* **4,** 1–7.
Takimoto, G., Tasset, D., Eppert, A., and Horwitz, K. (1992). Hormone-induced progesterone receptor phosphorylation consists of sequential DNA-independent and DNA-dependent stages: Analysis with zinc finger mutants and the progesterone antagonist ZK98299. *Proc. Natl. Acad. Sci. U.S.A.* **89,** 3050–3054.
Tasset, D., Tora, L., Fromental, C., Scheer, E., and Chambon, P. (1990). Distinct classes of transcriptional activating domains function by different mechanisms. *Cell* **62,** 1177–1187.
Tjian, R., and Maniatis, T. (1994). Transcriptional activation: A complex puzzle with a few easy pieces. *Cell* **77,** 5–8.
Tzukerman, M., Zhang, H. K., Hermann, T., Wills, K. N., Graupner, G., and Pfahl, M. (1990). The human estrogen receptor has transcriptional activator and repressor functions in the absence of ligand. *New Biol.* **2,** 613–620.
Tzukerman, M. T., Esty, A., Santiso-Mere, D., Danielian, P., Parker, M. G., Stein, R. B., Pike, J. W., and McDonnell, D. P. (1994). Human estrogen receptor transcriptional capacity is determined by both cellular and promoter context and mediated by two functionally distinct intramolecular regions. *Mol. Endocrinol.* **8,** 21–30.
Umesono, K., and Evans, R. M. (1989). Determinants of target gene specificity for steroid/thyroid hormone receptors. *Cell* **57,** 1139–1146.
Vegeto, E., Allan, G. F., Schrader, W. T., Tsai, M.-J., McDonnell, D. P., and O'Malley, B. W. (1992). The mechanism of RU486 antagonism is dependent on the conformation of the carboxy-terminal tail of the human progesterone receptor. *Cell* **69,** 703–713.
vonAngerer, E., Biberger, C., Holler, E., Koop, E., and Leichtl, S. (1994). 1-Car-

bamoylalkyl-2-phenylindoles: Relationship between side chain structure and estrogen antagonism. *J. Steroid Biochem. Mol. Biol.* **49,** 51–62.

Wagner, R. L., Apriletti, J. W., McGrath, M. E., West, B. L., Baxter, J. D., and Fletterick, R. J. (1995). A structural role for hormone in the thyroid hormone receptor. *Nature (London)* **378,** 690–697.

Wakeling, A. E., Dukes, M., and Bowler, J. (1991). A potent specific pure antiestrogen with clinical potential. *Cancer Res.* **51,** 3867–3873.

Wei, L. L., Krett, N. L., Francis, M. D., Gordon, D. F., Wood, W. M., O'Malley, B. W., and Horwitz, K. B. (1988). Multiple human progesterone receptor messenger ribonuclei acids and their autoregulation by progestin agonists and antagonists in breast cancer cells. *Mol. Endocrinol.* **2,** 62–72.

Wei, L. L., Leach, M. W., Miner, R. S., and Demers, L. M. (1993). Evidence for progesterone receptors in human osteoblast-like cells. *Biochem. Biophys. Res. Commun.* **195,** 525–532.

Weigel, N. L., Beck, C. A., Estes, P. A., Prendergast, P., Altman, M., Christensen, K., and Edwards, D. P. (1992). Ligands induce conformational changes in the carboxylterminus of progesterone receptors which are detected by a site directed antipeptide monoclonal antibody. *Mol. Endocrinol.* **6,** 1585–1597.

Wilson, A. C., LaMarco, K., Peterson, M. G., and Herr, W. (1993). The VP16 accessory protein HCF is a family of polypeptides processed from a large precursor protein. *Cell* **74,** 115–125.

Yu, L., Loewenstein, P. M., Zhang, Z., and Green, M. (1995). In vitro interaction of the human immunodeficiency virus type 1 Tat transactivator and the general transcription factor TFIIB with the cellular protein TAP. *J. Virol.* **69,** 3017–3023.

Zawel, L., Kumar, K. P., and Reinberg, D. (1995). Recycling of the general transcription factors during RNA polymerase II transcription. *Genes Dev.* **9,** 1479–1490.

Signal Transduction Pathways Combining Peptide Hormones and Steroidogenesis

MICHAEL R. WATERMAN AND DIANE S. KEENEY

Vanderbilt University School of Medicine, Department of Biochemistry, Nashville, Tennessee 37232

I. Introduction
II. Acute Action of Peptide Hormones on Steroidogenesis
III. Chronic Action of Peptide Hormones on Steroidogenesis
    A. Multifactorial Regulation of Steroidogenic Pathways
    B. Developmental–Tissue-Specific Regulation
    C. cAMP-Independent Regulation
    D. cAMP-Dependent Regulation
    E. Are Developmental–Tissue-Specific Regulation and cAMP-Dependent Regulation Coupled?
IV. Conclusions and Future Directions
References

I. Introduction

Peptide hormones released from the anterior pituitary bind to specific receptors on a limited number of cell types (steroidogenic cells). Signals resulting from this binding are amplified through the production of steroid hormones, leading to the regulation of transcription of genes in all cells. A major advancement in biology has been the identification and characterization of nuclear receptors that bind specific ligands, forming complexes that bind to specific DNA sequences through their zinc finger motifs and thereby regulating transcription of the associated genes. Levels of one class of ligands, the steroid hormones, are controlled by the action of peptide hormones from the anterior pituitary. Over the same period of time that the nuclear steroid hormone receptors have been characterized, an understanding of the regulatory processes leading to production of these steroidal ligands has emerged. Consequently, we now have a good view of how these peptide hormones exert their actions. Adrenocorticotropin (ACTH) receptors are found in the adrenal cortex, luteinizing hormone (LH) receptors in the testis and ovary, and follicle-stimulating hormone (FSH) receptors in the ovary. Each of these endocrine tissues is a factory for production of a specific subset of steroid hormones. In this way the endocrine roles of the adrenals and gonads serve to amplify the action of peptide hormones in a few cells to regulate gene transcription in all cells (Fig. 1).

Each of these peptide hormone receptors is coupled to adenylate cyclase through G-proteins, and consequently binding of the appropriate peptide hormone activates production of cAMP, which in turn activates steroid hormone biosynthesis by acute and chronic mechanisms. Both mechanisms are mediated by cAMP through cAMP-dependent protein kinase (PKA), and the aim of this article is to provide an overview of both actions of peptide hormones in the regulation of steroid hormone biosynthesis.

II. Acute Action of Peptide Hormones on Steroidogenesis

The general patterns for steroid hormone biosynthesis in humans and many other species are seen in Fig. 2. Briefly, endocrine amounts of aldosterone (mineralocorticoid), corticosterone, and cortisol (glu-

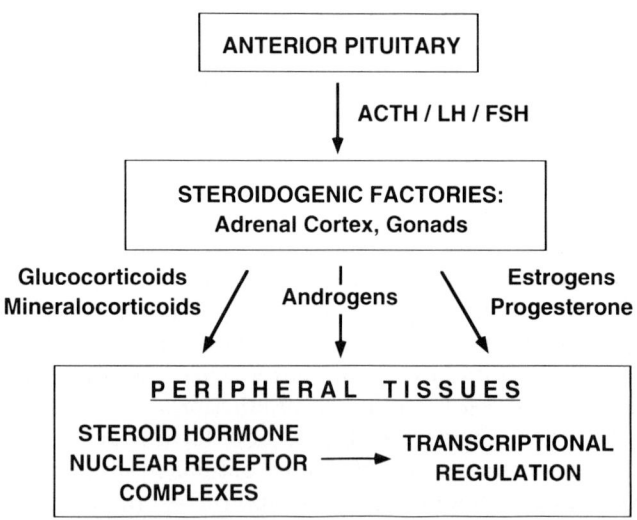

Fig. 1. Schematic representation of the pathway by which peptide hormones from the anterior pituitary regulate transcription of genes in all tissues. These peptide hormones activate steroid hormone biosynthesis in a limited number of sites, and these ligands for nuclear receptors find their way to all cells via the circulation.

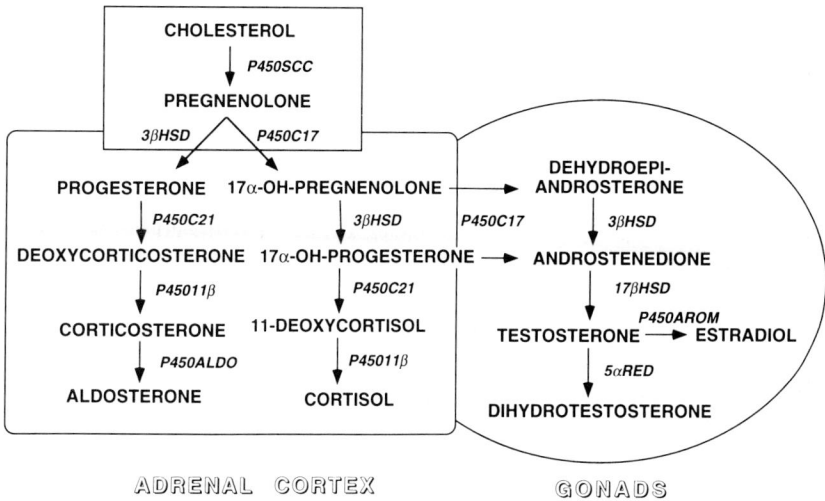

FIG. 2. Steroidogenic pathways found in most species. The first step, conversion of cholesterol to pregnenolone, occurs in all steroidogenic factories $P450_{scc}$, cholesterol side chain cleavage cytochrome P450; $P450_{c17}$, 17α-hydroxylase/17,20-lyase cytochrome P450; $P450_{c21}$, 21-hydroxylase cytochrome P450; $P450_{11\beta}$, 11β-hydroxylase cytochrome P450; $P450_{ALDO}$, aldosterone synthase; 3βHSD, 3β-hydroxysteroid dehydrogenase; 7βHSD, 7β-hydroxysteroid dehydrogenase, $P450_{AROM}$, aromatase cytochrome P450; 5αRED, 5α-reductase.

cocorticoids) are produced in the adrenal cortex; testosterone (male sex hormone) in the testis; and estrogen (female sex hormone) in the ovary. Progesterone, another important steroid hormone, is produced in the ovary and placenta. The placenta is also a factory for the production of steroid hormones; however, it appears that placental steroidogenesis is not regulated by peptide hormones through PKA in all species (Hum and Miller, 1993; Yamamoto et al., 1995). Each of the steroidogenic pathways begins in the mitochondrion, where cholesterol is converted to pregnenolone by cholesterol side chain cleavage cytochrome P450 ($P450_{scc}$), an integral protein of the inner mitochondrial membrane (Simpson, 1979). Mitochondrial membranes in all cells contain cholesterol as a structural component. Mitochondria in steroidogenic cells of the adrenal, testis, and ovary also contain a second pool of cholesterol that serves as substrate for $P450_{scc}$. cAMP regulates the level of this steroidogenic cholesterol pool in the inner mitochondrial membrane (Simpson and Waterman, 1995). It has long been known that peptide hormones can activate steroidogenesis via cAMP in a matter of minutes and that this process is inhibited by the protein synthesis inhibitor cycloheximide (Garren et al., 1971). It is presumed that a newly synthesized protein plays a key role in making available steroidogenic cholesterol in adrenals and gonads.

In the adrenal cortex, ACTH binding activates synthesis of cAMP (Richardson and Schulster, 1973; Wong and Schimmer, 1989). Through the action of PKA, cholesterol ester hydrolase is activated by phosphorylation. This enzyme cleaves the ester function from cholesterol esters, producing free cholesterol, which is the substrate for $P450_{scc}$ (Jefcoate et al., 1992). However, the next step, the transport of cholesterol within cells, is not at all well understood. It has been proposed that sterol carrier protein 2 (SCP2) participates in bringing free cholesterol from the lipid droplets, where it is produced in response to cAMP, to the mitochondrion (Chanderbhan et al., 1982). Also the level of SCP2 has been found to increase in rat adrenal cells in response to ACTH (Trzeciak et al., 1987). However, proof of this carrier role of SCP2 remains to be established. An alternative transport pathway for cholesterol from lipid droplets to mitochondria is through intermediate filaments of the cytoskeleton. The filaments are found by electron microscopy to be attached to both organelles (Almahbobi et al., 1992a,b) but the mechanism by which intermediate filaments could participate in cholesterol transport has not been established.

Once cholesterol arrives at the mitochondrion, it must find its way across the outer mitochondrial membrane and across the intermembrane space to the inner mitochondrial membrane where $P450_{scc}$ is located. Details of this movement of cholesterol from the surface of the mitochondrion to the vicinity of $P450_{scc}$ in the inner mitochondrial membrane are better understood than are the mechanisms of transport to the mitochondrial surface. Action of cAMP leads to accumulation of cholesterol in mitochondria in steroidogenic cells. In the presence of cycloheximide, this accumulation occurs only in the outer mitochondrial membrane (Privalle et al., 1983; Jefcoate et al., 1992). Thus a newly synthesized, labile protein factor is required for the movement of cholesterol from the outer to the inner mitochondrial membrane. This labile protein is thought to be present in the cells that serve as factories for the production of steroid hormones (the adrenal cortex and the gonads) but not in other cells where cholesterol transport occurs, such as the liver (Sugawara et al., 1995). Two-dimensional gel electrophoresis of proteins from Leydig cells has revealed proteins that appear rapidly following stimulation by cAMP (Epstein and

Orme-Johnson, 1991; Stocco and Sodeman, 1991). Only recently, however, has such a protein been identified by molecular cloning so that its biochemical function can be elucidated.

Clark and Stocco cloned a protein from Leydig cells that they have named *s*teroidogenic *a*cute *r*egulatory protein (StAR) (Clark et al., 1994). Overexpression of StAR in steroidogenic cells enhances production of the C21 steroid pregnenolone from the cholesterol, providing evidence for the role of StAR in enhancing steroidogenic activity (King et al., 1995). Further evidence for the role of StAR in making cholesterol available to $P450_{scc}$ in the inner membrane of mitochondria in steroidogenic cells comes from the study of congenital adrenal hyperplasia. The congenital adrenal hyperplasias are a group of genetic diseases involving many of the genes encoding the steroidogenic enzymes shown in Fig. 2 (White, 1994). The form known as congenital lipoid adrenal hyperplasia has also been called cholesterol side chain cleavage deficiency, and was thought until recently to result from mutations in the CYP11A ($P450_{scc}$) gene. However, unlike other forms of congenital adrenal hyperplasia that result from mutations in the exonic sequences of steroid hydroxylase genes, no such mutation has been found in the CYP11A gene in individuals suffering from congenital lipoid adrenal hyperplasia (Lin et al., 1991). Also, mitochondria from Leydig cells of an affected individual show functional $P450_{scc}$ activity (Sakai et al., 1994). Thus, the defect in congenital lipoid adrenal hyperplasia does not directly result from aberrant $P450_{scc}$. It has now been shown that congenital lipoid adrenal hyperplasia is associated with mutations in StAR, providing irrefutable evidence that this protein is essential for transport of cholesterol to $P450_{scc}$ (Lin et al., 1995).

The question remains, however, by what mechanism does StAR participate in the movement of cholesterol from the outer to the inner mitochondrial membrane? StAR could bind cholesterol or it could participate in formation of contact points between the outer and inner membranes that facilitate the passive flow of cholesterol down a gradient from outside to inside. This important mechanism remains to be established.

In summary, peptide hormones acting through cAMP provide for the rapid mobilization of cholesterol from cellular stores to the site of the first step in all steroidogenic pathways. This acute action of peptide hormones in the factories for steroid hormone production is an essential step in the overall scheme for peptide hormone action presented in Fig. 1.

III. Chronic Action of Peptide Hormones on Steroidogenesis

A. Multifactorial Regulation of Steroidogenic Pathways

The chronic response of steroidogenesis to peptide hormones results from transcriptional regulation of the genes encoding steroidogenic enzymes, which, like the acute response, is mediated by cAMP. Study of the chronic response of peptide hormone action has been focused on a few genes: CYP11A, CYP11B (P450$_{c11}$), CYP17 (P450$_{c17}$), CYP21 (P450$_{c21}$), 3βHSD (3β-hydroxysteroid dehydrogenase), and adrenodoxin. It is this action of peptide hormones (cAMP) that assures that optimal levels of steroidogenic enzymes are present in the factories that produce steroid hormones. Therefore, the acute and chronic responses of peptide hormones are linked, the former leading to activation of steroidogenesis by providing substrate for the pathway, the latter assuring that adequate levels of the enzymes required for biosynthesis of steroid hormones from cholesterol are always present. Before examining this chronic response, it is important to consider transcriptional regulation of steroidogenic enzymes in the broadest view. This regulation is multifactorial, involving developmental, tissue-specific, cAMP-independent, and cAMP-dependent mechanisms.

B. Developmental–Tissue-Specific Regulation

Developmental and tissue-specific regulation overlap one another and are not easily distinguished. Developmental regulation of steroidogenesis leads to the timely production of steroid hormones during fetal life. The most critical need for steroid hormones in development is the requirement for testosterone in establishing the male phenotype. The congenital adrenal hyperplasias provide an excellent view of the requirement for steroid hormones during development. In the absence of testosterone production, 46,XY (male) individuals develop a female phenotype, lacking male secondary sex characteristics (Yanase et al., 1991). Blockage of the synthesis of other steroid hormones during development does not lead to discernable defects in phenotype, although multiple metabolic defects result after birth if steroid hormone supplementation is not provided.

Developmental expression of steroidogenic enzymes in the adrenal and gonads requires a nuclear orphan receptor known as steroidogenic factor-1 (SF-1) (Ikeda et al., 1993), which has also been named Ad4BP (Honda et al., 1993). This transcription factor is expressed in a limited number of sites, including the adrenal cortex, the gonads, and the hypothalamus. Each of the steroid hydroxylase genes, and that encoding steroidogenic 3βHSD, contains one or more SF-1 response elements in its promoters (Morohashi et al., 1992). Temporally, SF-1 is expressed prior to the appearance of transcripts from steroid hydroxylase genes (Ikeda et al., 1994). Thus it is argued intuitively that SF-1 is required for steroid hydroxylase gene expression. This paradigm was tested by disrupting the gene encoding SF-1 (Luo et al., 1994). Surprisingly, the knockout mice were born without adrenals or gonads. Thus the role of SF-1 in triggering the developmental expression of enzymes required for steroidogenesis has not been directly tested. Nevertheless, it is perfectly reasonable to imagine that SF-1 is crucial for the development of steroidogenic pathways, in addition to its essential role in organogenesis of the factories that produce steroid hormones. While SF-1 is expressed very early in all of these factories and persists throughout life, it has a variable pattern of expression in the developing ovary. In this organ, it is detected early in embryonic development and then declines until after birth, when its level again increases (Ikeda et al., 1994). Thus steroid hormones produced by the ovary do not appear in significant quantities until after birth. Perhaps fetal production of estrogen would be deleterious to normal embryonic development.

The pattern of SF-1 expression and the presence of SF-1-binding sites in the promoters of all genes encoding steroidogenic enzymes suggest that SF-1 is the crucial transcription factor for both the developmental patterns and tissue specificity of steroidogenic pathways. However, steroidogenic capacity is not found in the hypothalamus, where SF-1 expression is also observed, and it has been established that steroidogenic enzymes are expressed in tissues that do not express SF-1. These sites include placenta (Yamamoto et al., 1995), certain regions of the brain (Shinoda et al., 1995), and embryonic gut (Keeney et al., 1995). In this latter site, it has been proven that steroidogenic genes are expressed in the SF-1 knockout mice. It is thought that in these sites, with the exception of the placenta, steroid hormone production is for local consumption (autocrine or paracrine). These results do, however, clearly establish that SF-1 is not required for expression of steroidogenic enzymes in all tissues. Perhaps there are two different mechanisms for tissue-specific expression of steroidogenic enzymes: one in the steroid hormone factories that involves SF-1 and one in sites of paracrine production of steroid hormones that is SF-1 independent. The placenta does not fit into either category, being a factory for progesterone synthesis but not expressing SF-1.

C. cAMP-Independent Regulation

At present, cAMP-independent transcription of steroidogenic enzymes has been studied in primary cell cultures and permanent cell lines. Cytokines and other factors are thought to play important roles in this level of regulation. cAMP-independent regulation was first identified from studies carried out in primary cultures of bovine adrenocortical cells. After several days of culture in the absence of ACTH or cAMP analogues, mRNA for one steroid hydroxylase ($P450_{c17}$) disappeared while mRNA for other members of the adrenocortical steroidogenic pathways ($P450_{scc}$, $P450_{c21}$, $P450_{11\beta}$, and adrenodoxin) remained, although at a lowered level (Zuber et al., 1986). This indicates that factors other than cAMP in the cell culture medium are responsible for maintaining this lowered level of mRNA for several bovine steroidogenic enzymes, and that expression of bovine CYP17 is absolutely dependent on cAMP. Subsequently specific cytokines and growth factors have been shown to regulate the levels of different steroidogenic enzymes in different cell culture systems (Orava et al., 1989; Perrin et al., 1990). It will take a great deal of effort to sort out the details of this complex cAMP-independent aspect of the multifactorial regulation of steroidogenic pathways.

D. cAMP-Dependent Regulation

cAMP-dependent regulation of steroidogenic enzymes is essential for the maintenance of steroidogenic pathways in the factories that produce steroid hormones. This level of regulation was first identified in studies using hypophysectomized rats. Removal of the pituitary led to dramatic reduction of steroidogenic enzymes in both the adrenal and the testis (Kimura, 1969; Purvis et al., 1973a,b). Pharmacological doses of peptide hormones increased the levels of steroidogenic enzymes in hypophysectomized rats. These peptide hormones function by elevating intracellular levels of cAMP. Later it was established that this action of cAMP is at the transcriptional level (John et al., 1986). Using cultured bovine adrenocortical cells, it was shown that levels of the steroidogenic enzymes, their activities, and their specific mRNAs are elevated in response to ACTH treatment (Waterman and Simpson, 1898). Subsequently it was shown by nuclear run-on experiments that this resulted from cAMP-dependent transcriptional activation of the genes encoding these enzymes (John et al., 1986). One candidate transcription system to regulate levels of steroidogenic enzymes was cAMP response element-binding protein/cAMP response element (CREB/

CRE), the widely distributed cAMP response system (Roesler et al., 1988). However, certain features of the cAMP responsiveness of steroidogenic enzymes suggested that CREB/CRE might not be the operative system. First, it takes several hours (4–6) for the cAMP response, as measured by increased steroid hydroxylase mRNA levels, to be manifest (John et al., 1986). Second, the response is inhibited by treatment with the protein synthesis inhibitor cycloheximide. The CREB/CRE system responds much more rapidly and is not affected by protein synthesis inhibitors. One other interesting feature of the system in bovine adrenocortical cells is that the increased transcription of steroidogenic genes is coordinate; the levels of mRNAs encoding $P450_{scc}$, $P450_{c11}$, $P450_{c21}$, $P450_{c17}$, and adrenodoxin all increased with the same temporal pattern (John et al., 1986). This result suggested that the same transcriptional system (common transcription factor and DNA element) was present in the genes of all the steroidogenic enzymes. The following working hypothesis was suggested as "promoter-bashing" studies were initiated on genes encoding steroidogenic enzymes:

$$\text{ACTH} \rightarrow \text{cAMP} \rightarrow \text{PKA} \rightarrow \begin{array}{c}\text{Transcription}\\ \text{factor X}\end{array} \rightarrow \begin{array}{c}\text{Coordinate transcription}\\ \text{of genes encoding}\\ \text{steroidogenic enzymes}\end{array}$$

Surprisingly, this hypothesis has been found to be incorrect. The biochemical mechanism supporting cAMP-dependent transcription of steroidogenic enzymes is far more complex than the simple scheme presented above. The first portion of this scheme seems to be correct for all steroidogenic genes. That is, peptide hormones from the anterior pituitary (ACTH, LH, FSH) bind to cell surface receptors that are found specifically associated with one of the factories for steroid hormone production. This binding activates adenylate cyclase through coupled G-protein, leading to elevated levels of intracellular cAMP. Through the use of Kin 8 cells, a form of mouse adrenal Y1 tumor cells having a mutation in the regulatory subunit of PKA so that this enzyme is essentially nonfunctional (Rae et al., 1979), it has been clearly established that PKA is necessary for cAMP-dependent transcription of steroidogenic enzymes. Thus, the biochemical mechanism for peptide hormone-dependent transcription of steroidogenic enzymes probably involves common steps up to the point of PKA action. Beyond this point, the mechanism is quite variable from gene to gene as well as in some cases from species to species (Waterman, 1994). The complexity is quite remarkable and unexpected, and its basis is not easy to explain.

Table I presents a summary of the cAMP response sequences (CRSs)

TABLE I
cAMP Response Systems in Bovine Steroid Hydroxylase Genes

Gene		cis-Element	Transactivator
CYP11A (P450$_{scc}$)	CRSI	^{118}ACTGAGTCTGGGAGGAGCT^{100}	Sp1 (zinc finger DNA-binding motif)
	CRSII	^{70}GACCGCCCTGTCAGCTTCTCA^{50}	Sp1
CYP11B1 (P450$_{11\beta}$)	CRE (Ad1)	^{67}TGACGTGA^{60}	CREB (leucine zipper DNA-binding motif)
	Ad4	^{331}CCAAGGTC^{324}	SF-1 (Ad4BP) (zinc finger DNA-binding motif)
			The Ad4 element enhances transactivation by the Ad1 element
CYP21 (P450$_{c21}$)	CRS	^{129}CTGTTTTGTGGGCGG^{115}	ASP (putative zinc finger DNA-binding motif)
CYP17 (P450$_{c17}$)	CRSI	^{243}TTGATGGACAGTGAGCAAG^{225}	Pbx1 + unknown partner (Pbx1 has a homeodomain DNA-binding motif)
	CRSII	^{76}TGAGCATTAACATAAAGTCAAGG-AGAAGGTCAGGGG^{40}	SF-1 vs COUP-TF (both have zinc finger DNA-binding motifs)

of bovine steroid hydroxylase genes, the species studied in greatest detail. It is apparent from promoter-bashing studies of different steroidogenic genes that different transcription factors (Table I) participate in peptide hormone-dependent transcription. With the exception of CREB binding to CRE in CYP11B1, it is not known whether direct phosphorylation of any of these transcription factors by PKA is required for their action. Why each gene should have its own unique cAMP-responsive system rather than the simpler paradigm is not apparent. This is particularly puzzling since developmental–tissue-specific regulation of all the steroidogenic genes in the steroid hormone factories involves the common transcription factor SF-1. During evolution, it seems that these genes have captured a common developmental–tissue-specific mechanism for regulating transcription, while each has captured a unique mechanism for regulation of cAMP-dependent transcription.

The species and organ in which the most detailed analyses of the biochemical mechanisms involved in cAMP-dependent transcription have been carried out is the bovine adrenal cortex. A summary of the cAMP-dependent transcription systems in these genes provides a view of the variability associated with this level of regulation. The bovine CYP11A gene ($P450_{scc}$) contains two CRS elements; each includes a binding site for the ubiquitous transcription factor Sp1 (Table I) (Ahlgren et al., 1990; Momoi et al., 1992; Venepally and Waterman, 1995). Both CRS elements bind only Sp1 and no other nuclear protein, and both participate in cAMP-dependent transcription of this gene. Sp1 is not generally thought to be involved in cAMP-dependent transcription. The hypothesis as to how it serves this role in regulation of CYP11A in bovine steroidogenic tissues is that Sp1 interacts with another nuclear protein. This complex then interacts with the basal transcription machinery (Venepally and Waterman, 1995). The hypothetical protein that interacts with Sp1 is not a DNA-binding protein. It is predicted to be expressed only in steroidogenic tissues, and it may be activated by PKA. Perhaps PKA activation facilitates its interaction with Sp1. Sp1 itself can be phosphorylated, but it is not known that PKA phosphorylates this transcription factor (Jackson et al., 1990). A biochemical system somewhat similar to this has been identified for CREB, which has been found to interact with a large nuclear protein (CBP) that is not a DNA-binding protein (Chrivia et al., 1993; Kwok et al., 1994). In the CREB/CBP system, however, it is CREB that is phosphorylated by PKA (Yamamoto et al., 1988). Both human and mouse CYP11A genes contain Sp1-binding sites at approximately the same positions in their promoters as in the bovine CYP11A gene. This suggests that the same mechanism might function in steroidogenic tissues in these species as well (Venepally and Waterman, 1995). Human CYP11A, however, also contains a CRE sequence far upstream from its promoter, and perhaps CREB also participates in cAMP-dependent transcription of the human gene (Inoue et al., 1991; Moore et al., 1990).

The bovine CYP11B1 (11β-hydroxylase) also contains a CRE element in its promoter. It has been established that this CREB-binding element is essential for cAMP-dependent transcription (Honda et al., 1990), and that it cooperates with a nearby Ad4BP (SF-1)-binding site for maximal cAMP-dependent transcription to be achieved (Morohashi et al., 1993). The details of the interaction between CREB and SF-1 remain to be established.

The bovine CYP21 gene also contains an Sp1-binding site closely associated with its CRS (Kagawa and Waterman, 1991). However, binding of Sp1 is not required for cAMP-dependent transcription. Rather, a 78-kDa protein named ASP, whose binding site overlaps with the Sp1 site, is the key cAMP-dependent transcription factor for this gene (Kagawa and Waterman, 1992). ASP has not yet been cloned, so its structure and potential insight into its function remain unknown. Because the ASP- and Sp1-binding sites overlap, it will be interesting to determine whether binding of Sp1 represses cAMP-dependent transcription of CYP21 as a contrast to activation by ASP. If this were true, the function of Sp1 binding near the CYP21 CRS would be just the opposite of its function in the CYP11A CRS elements. Human CYP21 contains essentially the same CRS as bovine CYP21. Overlapping ASP- and Sp1-binding sites are present in approximately the same position in the promoter, and the ASP site is required for cAMP responsiveness (Kagawa and Waterman, 1991).

The bovine CYP17 (17α-hydroxylase) gene contains two CRS elements (Lund et al., 1990) that bind very different transcription factors. The proteins binding to CRSI have been purified from nuclear extracts using DNA affinity chromatography. Four DNA-binding proteins were purified using this upstream CRS (Zanger et al., 1991; Kagawa et al., 1994). A 43- and a 60-kDa protein were purified in much higher amounts than two 53-kDa proteins. The 43-kDa protein and one 53-kDa protein were identified as members of the PBX gene family (Pbx1b and Pbx1a, respectively), which has a homeodomain DNA-binding motif. PBX was discovered as part of a chromosomal translocation (t1:19) involved in pre-β cell acute lymphoblastic leukemia (Kamps et al., 1990). The fusion protein contains the transactivation domain of the transcription factor E2A and the DNA-binding region of

Pbx. Pbx is a homeodomain protein that is closely related to the *Drosophila* homeodomain protein extradenticle (Rauskolb and Wieschaus, 1994). CYP17 CRSI is the first endogenous Pbx-binding site to be discovered. Pbx1 contains a potential PKA phosphorylation site but it has not yet been proven that it is phosphorylated during its function in CYP17 cAMP-dependent transcription. The second major CRSI-binding protein (60 kDa) has not yet been cloned, nor has the other 53-kDa protein. From a variety of preliminary results and the fact that extradenticle functions to regulate *Drosophila* gene expression only with a partner protein (Chan *et al.*, 1994), it is predicted that Pbx1b forms a heterodimer with the 60-kDa protein that functions to regulate cAMP-dependent transcription of bovine CYP17. One intriguing observation is that, while the cDNA sequences for mouse and human Pbx1b are 95% identical, their amino acid sequences (243 amino acids) are 100% identical (Kagawa *et al.*, 1994).

The CRSII of bovine CYP17 binds two transcription factors, SF-1 and chicken ovalbumin upstream promoter transcription factor (COUP-TF) (Bakke and Lund, 1994). Both have zinc finger DNA-binding motifs and compete with one another for the same binding site. It has been found that SF-1 is a positive regulator of CYP17 transcription, while COUP-TF represses transcription. The mechanism of this yin–yang regulatory process remains to be established.

In summary, it is apparent that peptide hormone-dependent transcription of steroid hydroxylase genes in bovine adrenal cortex involves a unique biochemical system for each gene (Waterman, 1994). These range from the well known CREB/CRE system to a novel role for the ubiquitous transcription factor Sp1, to a heterodimer involving Pbx (a widely distributed homeodomain protein), to a novel transcription factor predominantly expressed in the adrenal cortex (ASP). Also, SF-1, the orphan receptor that plays an essential role in developmental–tissue-specific transcription of all these genes, appears to participate in cAMP-dependent transcription of bovine CYP11B1 and CYP17. As indicated previously, the peptide hormone-dependent transcription of these genes occurs in a coordinate fashion. The reason for such a complex array of biochemical mechanisms leading to coordinated transcriptional events is absolutely unclear, especially since the developmental–tissue-specific transcription of these genes involves a common biochemical system (SF-1). It is predicted that all bovine steroidogenic tissues (adrenal cortex, testis, ovary) utilize the same biochemical mechanisms outlined earlier for their respective steroid hydroxylase genes. In comparing the sketchy results from different species, it appears that some genes (i.e., human and bovine CYP21) utilize very similar or identical biochemical mechanisms while others (CYP17) may involve quite different biochemical systems. Again, the reason for the variability in certain genes between different species is not obvious.

E. ARE DEVELOPMENTAL–TISSUE-SPECIFIC REGULATION AND cAMP-DEPENDENT REGULATION COUPLED?

SF-1 functions in both developmental–tissue-specific and cAMP-dependent transcription. Perhaps, then, peptide hormone-dependent transcription is important in fetal life as well as in postnatal and adult life. In one instance this clearly seems to be the case. Among the bovine steroidogenic enzymes, CYP17 is unique in that it is regulated exclusively by cAMP; it shows no cAMP-independent regulation. Therefore, in primary bovine adrenocortical cells maintained in the absence of peptide hormones, mRNA encoding this enzyme disappears after only a few days of culture in the absence of ACTH (Zuber et al., 1986). The fetal bovine adrenal cortex expresses CYP17 early in fetal life. However, during the midtrimester of fetal development, CYP17 expression is not observed (Lund et al., 1988). Expression reappears in the last trimester. By contrast, expression of CYP11A in the fetal adrenal and of CYP17 in fetal testis persists throughout fetal life. From a physiological standpoint, this result indicates that fetal cortisol production cannot occur during the midtrimester of development. This is confirmed by steroid hormone measurements in fetal blood. Corticosterone was readily detected but no cortisol was found during the midtrimester of fetal development. Measurement of ACTH levels in bovine fetal blood correlate well with levels of CYP17: both are undetectable in the midtrimester. This confirms the absolute requirement for peptide hormone-dependent generation of cAMP for bovine CYP17 transcription. It also suggests that the cAMP-dependent regulatory mechanisms required for steroid hydroxylase expression in the adult adrenal are also essential in the fetus. Thus we can predict that, while SF-1 is essential for the initial expression of steroid hydroxylases in steroid hormone factories, peptide hormones are essential for continued steroid hydroxylase expression during fetal life. Both SF-1 and cAMP-dependent mechanisms are necessary for fetal steroidogenic factories to have the capacity to produce steroid hormones.

The role of steroid hormones produced during fetal life is not clearly established. One absolutely essential role is the necessity of testosterone for development of the male phenotype. Thus, it can be predicted that LH-dependent regulation of the levels of steroidogenic enzymes involved in testosterone biosynthesis in the testis is crucial during development. All SF-1 knockout mice, which have no steroidogenic factories, have a female phenotype. However, they are born and breath before they expire. Therefore, the precise role of steroid hormones produced in the fetus (with the exception of testosterone) remains uncertain. It can be argued that steroid hormones from the mother are sufficient for developmental needs of the fetus. However, one other intriguing aspect of the studies on the bovine fetus is that no cortisol is detected in fetal blood when ACTH is not present. Thus in the cow, even though the mother produces cortisol, this steroid hormone is not transferred through the placenta to the fetus. If we can assume that maternal steroids are not transferred to the SF-1 knockout mouse, then steroid hormones produced by fetal steroidogenic factories are not necessary for development. This might explain why certain steroid hydroxylases are expressed in nonsteroidogenic tissues in an SF-1–independent fashion. Perhaps steroid hormones produced locally play important developmental roles and steroid hormones from factories, with the exception of testosterone, do not.

While SF-1 is essential to trigger the initial appearance of steroid hormone biosynthetic pathways, peptide hormone-dependent regulation is required for establishing optimal steroidogenic capacity from the very earliest development of the organs that serve as factories for steroid hormones. Since adrenocortical $P450_{scc}$ is present in the midtrimester of bovine fetal life when ACTH is absent, it would seem that cAMP-independent regulation of transcription of steroidogenic enzymes is also necessary very early in fetal life. Thus all of the multifactorial aspects of transcriptional regulation of steroidogenic pathways seem to function early in fetal life, including that controlled by peptide hormones.

IV. Conclusions and Future Directions

As outlined schematically in Fig. 1, the primary function of peptide hormones from the anterior pituitary is to regulate expression of a wide variety of different genes in many or perhaps all tissues. They are able to serve this role by regulating the production of steroid hormones in a small number of factories (adrenal and gonads). These steroid hormones pass through the circulation to all tissues, where they bind their respective nuclear receptors for transcriptional regulation. The action of peptide hormones in these factories is two-pronged, but both mechanisms depend on the same initial events resulting from peptide hormones binding to their specific receptors. In all cases this binding of peptide hormones activates adenylate cyclase through G-proteins, leading to elevated levels of intracellular cAMP.

In a rapid series of events referred to as the acute response of peptide hormones, cAMP activates the mobilization of cholesterol from cellular lipid stores into the inner mitochondrial membrane, where P450$_{scc}$ (the first enzyme in all steroidogenic pathways) resides. This mobilization involves elevation of levels of StAR, which plays an essential, yet to be defined, role in mobilizing cholesterol from the outer mitochondrial membrane across the intermembrane space to the inner mitochondrial membrane. Thus, steroid hormone production is rapidly activated in response to elevated cAMP levels due to increased availability of substrate.

A slower response to elevated levels of intracellular cAMP, known as the chronic response to peptide hormones, activates transcription of genes encoding enzymes required for steroid hormone biosynthesis. The chronic response assures that optimal steroidogenic capacity is maintained in the factories producing steroid hormones. Although the levels of steroidogenic enzymes are elevated coordinately in response to peptide hormones, it has been surprising to find that the biochemical mechanisms necessary for enhanced transcription differ among steroidogenic genes. Furthermore, coupling of this chronic response with developmental regulation of steroid hydroxylase gene expression indicates the importance of the chronic response in maintaining optimal steroidogenic capacity during fetal life.

Several key aspects of the acute and chronic actions of peptide hormones remain to be clarified. For example, how is cholesterol transported to the mitochondrion, and what precise role does StAR play in getting cholesterol to the inner mitochondrial membrane? Both the acute and chronic responses are mediated via PKA, but why does the chronic response take several hours to become manifest? Is the action of PKA an early event in a multistep process that does not include subsequent cAMP-dependent phosphorylation but nevertheless leads to activated transcription of these genes? In fact, because each gene utilizes its own biochemical mechanism for activating transcription, is the early action of PKA in this process the last common event for coordinated transcription? Finally, the details of how each unique transcription system interacts with the basal transcription machinery must be established. We have come a long way in deciphering how peptide hormones regulate the synthesis of steroid hormones and subsequently the transcription of many different genes in all tissues via nuclear receptors. However, each step in elucidating the mechanisms by which peptide hormones regulate steroid hormone levels has led to new questions, and we are far short of having a complete understanding of this action of peptide hormones.

ACKNOWLEDGMENTS

The authors acknowledge the accomplishments of their many colleagues who have aided in their understanding of how peptide hormones regulate steroid hormone synthesis. Of particular note is Dr. Evan R. Simpson. The opportunity to study this fascinating system provided by U.S. Public Health Service grant DK28350 is also acknowledged.

REFERENCES

Ahlgren, R., Simpson, E. R., Waterman, M. R., and Lund, J. (1990). Characterization of the promoter-regulatory region of the bovine CYP11A (P450$_{scc}$) gene: Basal and cAMP-dependent expression. *J. Biol. Chem.* **265**, 3313–3319.

Almahbobi, G., Williams, L. J., and Hall, P. F. (1992). Attachment of mitochondria to intermediate filaments in adrenal cells: Relevance to the regulation of steroid synthesis. *Exp. Cell Res.* **200**, 361–369.

Almahbobi, G., Williams, L. J., and Hall, P. F. (1992). Attachment of steroidogenic lipid droplets to intermediate filaments in adrenal cells. *J. Cell Sci.* **101**, 383–393.

Bakke, M., and Lund, J. (1994). Mutually exclusive interactions of two nuclear orphan receptors determine activity of a cAMP-responsive sequence in the bovine CYP17 gene. *Mol. Endocrinol.* **9**, 327–339.

Chan, S.-K., Jaffe, L., Capovilla, M., Botas, J., and Mann, R. S. (1994). The DNA binding specificity of ultrabithorax is modulated by cooperative interactions with extradenticle, another homeoprotein. *Cell* **78**, 603–615.

Chanderbhan, R., Noland, B. J., Scallen, T. J., and Vahouny, G. V. (1982). Sterol carrier proteins$_2$: Delivery of cholesterol from adrenal lipid droplets to mitochondria for pregnenolone synthesis. *J. Biol. Chem.* **257**, 8928–8934.

Chrivia, J. C., Kwok, R. P. S., Lamb, N., Hagiwara, M., Montminy, M. R., and Goodman, R. H. (1993). Phosphorylated CREB binds specifically to the nuclear protein CBP. *Nature (London)* **365**, 855–859.

Clark, B. J., Wells, J., King, S. R., and Stocco, D. M. (1994). The purification, cloning, and expression of a novel luteinizing hormone-induced mitochondrial protein in MA-10 mouse Leydig tumor cells. *J. Biol. Chem.* **269**, 28318–28322.

Epstein, L. F., and Orme-Johnson, N. R. (1991). Regulation of steroid hormone biosynthesis: Identification of precursors of a phosphoprotein targeted to the mitochondrion in stimulated rat adrenal cortex cells. *J. Biol. Chem.* **266**, 19739–19745.

Garren, L. D., Gill, G. N., Masui, H., and Walton, G. M. (1971). On the mechanism of action of ACTH. *Recent Prog. Horm. Res.* **27**, 433–478.

Honda, S., Morohashi, K., and Omura, T. (1990). Novel cAMP regulatory elements in the promoter region of bovine P45011β gene. *J. Biochem.* **108**, 1042–1049.

Honda, S., Morohashi, K., Nomura, M., Takeya, M., Kitajimi, M., and Omura, T. (1993). Ad4BP regulating steroidogenic P450 genes is a member of steroid hormone receptor superfamily. *J. Biol. Chem.* **268**, 7479–7502.

Hum, D. W., and Miller, W. L. (1993). Transcriptional regulation of human genes for steroidogenic enzymes. *Clin. Chem.* **39**, 333–340.

Ikeda, Y., Lala, D. S., Luo, X., Kim, E., Moisan, M.-P., and Parker, K. L. (1993). Charac-

terization of the mouse FTZ-F1 gene, which encodes a key regulator of steroid hydroxylase gene expression. *Mol. Endocrinol.* **7,** 852–860.
Ikeda, Y., Shen, W.-H., Ingraham, H. A., and Parker, K. L. (1994). Developmental expression of mouse steroidogenic factor-1, an essential regulator of the steroid hydroxylases. *Mol. Endocrinol.* **8,** 654–662.
Inoue, H., Watanabe, N., Higashi, Y., and Fujii-Kuriyama, Y. (1991). Structures of regulatory regions in the human cytochrome P450$_{scc}$ (desmolase) gene. *Eur. J. Biochem.* **195,** 563–569.
Jackson, S. P., MacDonald, J. J., Lees-Miller, S., and Tjian, R. (1990). GC box binding induces phosphorylation of Sp1 by a DNA-dependent protein kinase. *Cell* **63,** 155–165.
Jefcoate, C. R., McNamara, B. C., Artemenko, I., and Yamazaki, T. (1992). Regulation of cholesterol movement to mitochondrial cytochrome P450$_{scc}$ in steroid hormone synthesis. *J. Steroid Biochem. Mol. Biol.* **43,** 751–767.
John, M. E., John, M. C., Boggaram, V., Simpson, E. R., and Waterman, M. R. (1986). Transcriptional regulation of steroid hydroxylase genes by ACTH. *Proc. Natl. Acad. Sci. U.S.A.* **83,** 4715–4719.
Kagawa, N., and Waterman, M. R. (1991). Evidence that an adrenal-specific nuclear protein regulates cAMP responsiveness of the human CYP21B (P450$_{C21}$) gene. *J. Biol. Chem.* **266,** 11199–11204.
Kagawa, N., and Waterman, M. R. (1992). Purification and characterization of a transcription factor which appears to regulate the cAMP-responsiveness of the human CYP21B gene. *J. Biol. Chem.* **267,** 25213–25219.
Kagawa, N., Ogo, A., Takahashi, Y., Iwamatsu, A., and Waterman, M. R. (1994). A cAMP-responsive sequence (CRS1) of CYP17 is a cellular target for the homeodomain protein Pbx1. *J. Biol. Chem.* **269,** 18716–18719.
Kamps, M. P., Murre, C., Sun, X. H., and Baltimore, D. (1990). A new homeobox gene contributes the DNA binding domain of the t(1;19) translocation protein in pre-B all. *Cell* **60,** 547–555.
Keeney, D. S., Ikeda, Y., Waterman, M. R., and Parker, K. (1995). Cholesterol side-chain cleavage cytochrome P450 gene expression in primitive gut of the mouse embryo does not require steroidogenic factor-1. *Mol. Endocrinol.* **19,** 1091–1098.
Kimura, T. (1969). Effects of hypophysectomy and ACTH administration on the level of adrenal cholesterol side-chain desmolase. *Endocrinology* **85,** 492–499.
King, S. R., Ronen-Fuhrmann, T., Timberg, R., Clark, B. J., Orly, J., and Stocco, D. M. (1995). Steroid production after *in vitro* transcription, translation, and mitochondrial processing of protein products of complementary deoxyribonucleic acid for steroidogenic acute regulatory protein. *Endocrinology* **136,** 5165–5176.
Kwok, R. P. S., Lundblad, J. R., Chrivia, J. C., Richards, J. P., Bächinger, H. P., Brennan, R. G., Roberts, S. G. E., Green, M. R., and Goodman, R. H. (1994). CBP serves as a coactivator for the transcription factor CREB. *Nature (London)* **370,** 223–229.
Lin, D., Gitelman, S. E., Saenger, P., and Miller, W. L. (1991). Normal genes for cholesterol side chain cleavage enzyme, P450scc, in congenital lipoid adrenal hyperplasia. *J. Clin. Invest.* **88,** 1955–1962.
Lin, D., Sugawara, T., Strauss, J. F., Clark, B. J., Stocco, D. M., Saenger, P., Rogol, A., and Miller, W. L. (1995). Role of steroidogenic acute regulatory protein in adrenal and gonadal steroidogenesis. *Science* **267,** 1828–1831.
Lund, J., Faucher, D. J., Ford, S. P., Porter, J. C., Waterman, M. R., and Mason, J. I. (1988). Developmental expression of bovine adrenocortical steroid hydroxylases: Regulation of P-450$_{17\alpha}$ expression leads to episodic fetal cortisol production. *J. Biol. Chem.* **263,** 16195–16201.

Lund, J., Ahlgren, R., Wu, D., Kagimoto, M., Simpson, E. R., and Waterman, M. R. (1990). Transcriptional regulation of the bovine CYP17 (P450$_{17\alpha}$) gene: Identification of two cAMP regulatory regions lacing the consensus CRE. *J. Biol. Chem.* **265**, 3304–3312.

Luo, X., Ikeda, Y., and Parker, K. L. (1994). A cell-specific nuclear receptor is essential for adrenal and gonadal development and sexual differentiation. *Cell* **77**, 481–490.

Momoi, K., Waterman, M. R., Simpson, E. R., and Zanger, U. M. (1992). 3′, 5′-cyclic adenosine monophosphate-dependent transcription of the CYP11A (cholesterol side chain cleavage cytochrome P450) gene involves a DNA response element containing a putative binding site for transcription factor Sp1. *Mol. Endocrinol.* **6**, 1682–1690.

Moore, C. C. D., Brentano, S. T., and Miller, W. L. (1990). Human P450scc gene transcription is induced by cAMP and repressed by 1200-tetradecanoylphorbol-13-acetate and A23187 through independent *cis*-elements. *Mol. Cell. Biol.* **10**, 6013–6023.

Morohashi, K., Honda, S., Inomata, Y., Handa, H., and Omura, T. (1992). A common *trans*-acting factor, Ad4-binding protein, to the promoters of steroidogenic P450s. *J. Biol. Chem.* **267**, 17913–17919.

Morohashi, K., Zanger, U. M., Honda, S., Hara, M., Waterman, M. R., and Omura, T. (1993). Activation of CYP11A and CYP11B gene promoters by the steroidogenic cell-specific transcription factor, Ad4BP. *Mol. Endocrinol.* **7**, 1196–1204.

Orava, M., Voutilainen, R., and Vihko, R. (1989). Interferon-γ inhibits steroidogenesis and accumulation of mRNA of the steroidogenic enzymes P450scc and P450c17 in cultured porcine Leydig cells. *Mol. Endocrinol.* **3**, 887–894.

Perrin, A., Pascal, O., Defaye, G., Feige, J.-J., and Chambaz, E. M. (1990). Transforming growth factor β is a negative regulatory of steroid 17α-hydroxylase expression in bovine adrenocortical cells. *Endocrinology* **128**, 357–362.

Privalle, C. T., Crivello, J. F., and Jefcoate, C. R. (1983). Regulation of intramitochondrial cholesterol transfer to side-chain cleavage cytochrome P450 in rat adrenal gland. *Proc. Natl. Acad. Sci. U.S.A.* **80**, 702–706.

Purvis, J. L., Canick, J. A., Latif, S. A., Rosenbaum, J. H., Hologgitar, J., and Menard, R. H. (1973a). Lifetime of microsomal cytochrome P450 and steroidogenic enzymes in rat testis as influenced by human chorionic gonadotropin. *Arch. Biochem. Biophys.* **159**, 39–49.

Purvis, J. L., Canick, J. A., Mason, J. I., Estabrook, R. W., and McCarthy, J. L. (1973b). Lifetime of adrenal cytochrome P450 as influenced by ACTH. *Ann. N.Y. Acad. Sci.* **212**, 319–342.

Rae, P. A., Gutmann, N. S., Tsao, J., and Schimmer, B. P. (1979). Mutations in cyclic AMP-dependent protein kinase and corticotropin (ACTH)-sensitive adenylate cyclase affect adrenal steroidogenesis. *Proc. Natl. Acad. Sci. U.S.A.* **76**, 1896–1900.

Rauskolb, C., and Wieschaus, E. (1994). Coordinate regulation of downstream genes by extradenticle and the homeotic selector proteins. *EMBO J.* **13**, 3561–3569.

Richardson, M. C., and Schulster, D. (1973). The role of protein kinase activation in the control of steroidogenesis in the adrenal cortex. *Biochem. J.* **136**, 993–998.

Roesler, W. J., Vandenbar, G. R., and Hanson, R. W. (1988). Cyclic AMP and the induction of eucaryotic gene transcription. *J. Biol. Chem.* **263**, 9063–9066.

Sakai, Y., Yanase, T., Okabe, Y., Hara, T., Waterman, M. R., Takayanagi, R., Haji, J., and Nawata, N. (1994). No mutation in cytochrome P450$_{scc}$ in a patient with congenital lipoid adrenal hyperplasia. *J. Clin. Endocrinol. Metab.* **79**, 1198–1201.

Shinoda, K., Lei, H., Yoshii, H., Nomura, M., Nagano, M., Shiba, H., Sasaki, H., Osawa, Y., Ninomiya, Y., Niwa, O., Morohashi, K.-I., and Li, E. (1995). Developmental defects of the ventromedial hypothalamic nucleus and pituitary gonadotroph in the *Ftz-F1* disrupted mice. *Dev. Dyn.* **204**, 22–29.

Simpson, E. R. (1979). Cholesterol side-chain cleavage cytochrome P450 and the control of steroidogenesis. *Mol. Cell. Endocrinol.* **13,** 213–227.

Simpson, E. R., and Waterman, M. R. (1995). Steroid hormone biosynthesis in the adrenal cortex and its regulation by adrencorticotropin. *In* "Endocrinology" (L. J. DeGroot, ed.), 3rd ed., pp. 1630–1641. Grune & Stratton, New York.

Stocco, D. M., and Sodeman, T. C. (1991). The 30 kDa mitochondrial proteins induced by hormonal stimulation in MA-10 mouse Leydig tumor cells are processed from larger precursors. *J. Biol. Chem.* **266,** 19731–19738.

Sugawara, T., Holt, J. A., Driscoll, D., Strauss, J. F. III, Lin, D., Miller, W. L., Patterson, D., Clancy, K. P., Hart, I. M., Clark, B. J., and Stocco, D. M. (1995). Human steroidogenic acute regulatory protein: Functional activity in COS-1 cells, tissue-specific expression, and mapping of the gene to 8p11.2 and a pseudogene to chromosome 13. *Proc. Natl. Acad. Sci. U.S.A.* **92,** 4778–4782.

Trzeciak, W. H., Simpson, E. R., Scallen, T. J., Vahouny, G. V., and Waterman, M. R. (1987). Studies on the synthesis of sterol carrier protein-2 in rat adrenocortical cells in monolayer culture: Regulation by ACTH and dibutyryl cyclic 3', 5' AMP. *J. Biol. Chem.* **262,** 3713–3717.

Venepally, P., and Waterman, M. R. (1995). Two Sp1-binding sites mediate cAMP-induced transcription of the bovine CYP11A gene through the protein kinase a signaling pathway. *J. Biol. Chem.* **270,** 25402–25410.

Waterman, M. R. (1994). Biochemical diversity of cAMP-dependent transcription of steroid hydroxylase genes in the adrenal cortex. *J. Biol. Chem.* **269,** 27783–27786.

Waterman, M. R., and Simpson, E. R. (1989). Regulation of steroid hydroxylase gene expression is multifactorial in nature. *Recent Prog. Horm. Res.* **45,** 533–566.

White, P. C. (1994). Genetic diseases of steroid metabolism. *Vitam. Horm.* **49,** 131–195.

Wong, M., and Schimmer, B. P. (1989). Recovery of responsiveness to ACTH and cAMP in a protein kinase-defective adrenal cell mutant following transfection with a protein kinase A gene. *Endoc. Res.* **15,** 49–65.

Yamamoto, K. K., Gonzelez, G. A., Biggs, W. H. III, and Montminy, M. R. (1988). Phosphorylation-induced binding and transcriptional efficacy of nuclear factor CREB. *Nature (London)* **334,** 494–498.

Yamamoto, T., Chapman, B. M., Clemens, J. W., Richards, J. A. S., and Soares, M. J. (1995). Analysis of cytochrome P450 side-chain cleavage gene promoter activation during trophoblast cell differentiation. *Mol. Cell. Endocrinol.* **113,** 183–194.

Yanase, T., Simpson, E. R., and Waterman, M. R. (1991). 17α-Hydroxylase/17,10-lyase deficiency: From clinical investigation to molecular definition. *Endocr. Rev.* **12,** 91–108.

Zanger, U. M., Lund, J., Simpson, E. R., and Waterman, M. R. (1991). Activation of transcription in cell-free extracts by a novel cAMP responsive sequence from the bovine CYP17 gene. *J. Biol. Chem.* **266,** 11417–11420.

Zuber, M. X., John, M. E., Okamura, T., Simpson, E. R., and Waterman, M. R. (1986). Bovine adrenocortical cytochrome P45017α: Regulation of gene expression by ACTH and elucidation of primary sequence. *J. Biol. Chem.* **261,** 2475–2482.

The Roles of 14-3-3 Proteins in Signal Transduction

GARY W. REUTHER AND ANN MARIE PENDERGAST

Department of Pharmacology, Duke University Medical Center, Durham, North Carolina 27710

I. Introduction
II. 14-3-3 Sequence and Structure Analysis
III. Regulation of Cellular Processes by 14-3-3 Proteins
   A. Catecholamine Biosynthesis
   B. Exocytosis
   C. ADP Ribosylation
   D. Cell Cycle
   E. Other Activities
IV. 14-3-3 Proteins and Protein Kinases
   A. Protein Kinase C
   B. Raf
   C. Bcr and Bcr-Abl
V. Roles of 14-3-3 Proteins in Signal Transduction Pathways
   A. Chaperones
   B. Molecular Bridges
   C. Effectors
VI. Conclusions
   References

I. Introduction

14-3-3 proteins constitute a highly conserved family of eukaryotic proteins (Aitken *et al.*, 1992). While proteins of this family were first described in 1967, their biological significance has remained elusive. Recent research in the field of signal transduction has elicited a great interest in determining the role(s) of 14-3-3 proteins within the cell. 14-3-3 proteins have been shown to regulate protein kinase C (PKC) and interact with middle T antigen, the Raf serine kinase, the breakpoint cluster region (Bcr) serine kinase, the Bcr-Abelson leukemia virus (Abl) tyrosine kinase, and the Cdc25 phosphatase, among other proteins (Freed *et al.*, 1994; Pallas *et al.*, 1994; Fu *et al.*, 1994; Irie *et al.*, 1994; Fantl *et al.*, 1994; Li *et al.*, 1995; Reuther *et al.*, 1994; Yamamori *et al.*, 1995).

The 14-3-3 protein family was first described during an analysis of brain proteins by Moore and Perez (1967). The name "14-3-3" comes from the nomenclature used by these researchers as they analyzed proteins by chromatography and electrophoresis. These proteins have a molecular mass of approximately 28–30 kDa, are acidic with isoelectric points of 4–5, and are dimeric in nature (Moore and Perez, 1968; Boston et al., 1982a,b; Toker et al., 1990, 1992). At first it was thought that 14-3-3 proteins were only present in the brain (Moore and Perez, 1967). To date these proteins have been found in all cell types examined, with highest expression in the brain (Aitken et al., 1992). 14-3-3 proteins are not restricted to vertebrate species. 14-3-3 protein homologues have been found in numerous organisms, including yeast, plants, the fruit fly *Drosophila melanogaster,* and the nematode *Caenorhabditis elegans* (Ford et al., 1994; van Heusden et al., 1992, 1995; Swanson and Ganguly, 1992; Wang and Shakes, 1994; Aitken et al., 1992; Aitken, 1995). The strong conservation of 14-3-3 proteins across species suggests that these proteins may play important roles in eukaryotic cells.

A large number of biochemical activities have been ascribed to 14-3-3 proteins, including regulation of hydroxylase activity in neurotransmitter biosynthesis, stimulation of secretion, ADP ribosylation cofactor activity, phospholipase A_2 activity, and regulation of PKC activity (Aitken et al., 1992; Aitken, 1995; Burbelo and Hall, 1995). This article describes the current understanding of 14-3-3 proteins. Speculations on the role of 14-3-3 proteins within the cell are presented.

II. 14-3-3 Sequence and Structure Analysis

The growing family of 14-3-3 proteins contains at least seven human gene products, named by Greek letters, which are highly identical at the amino acid level (Aitken et al., 1992). Comparison of human isoforms to 14-3-3 proteins of other eukaryotic species reveals a high level of homology. Figure 1 presents an alignment of three human 14-3-3 proteins with isoforms found in other species. This alignment shows that the homology among isoforms is widespread throughout the molecule. Several regions of the protein have an extremely high level of identity among most isoforms. The primary structure of 14-3-3 proteins predicts a highly α-helical structure (Toker et al., 1992). Regions of 14-3-3 have homology to the PKC pseudosubstrate sequence and to a group of proteins called the annexins (Fig. 1) (Aitken et al., 1992). While the carboxyl terminus is not highly conserved, it contains a large number of acidic residues. The significance of these regions is discussed in the following sections.

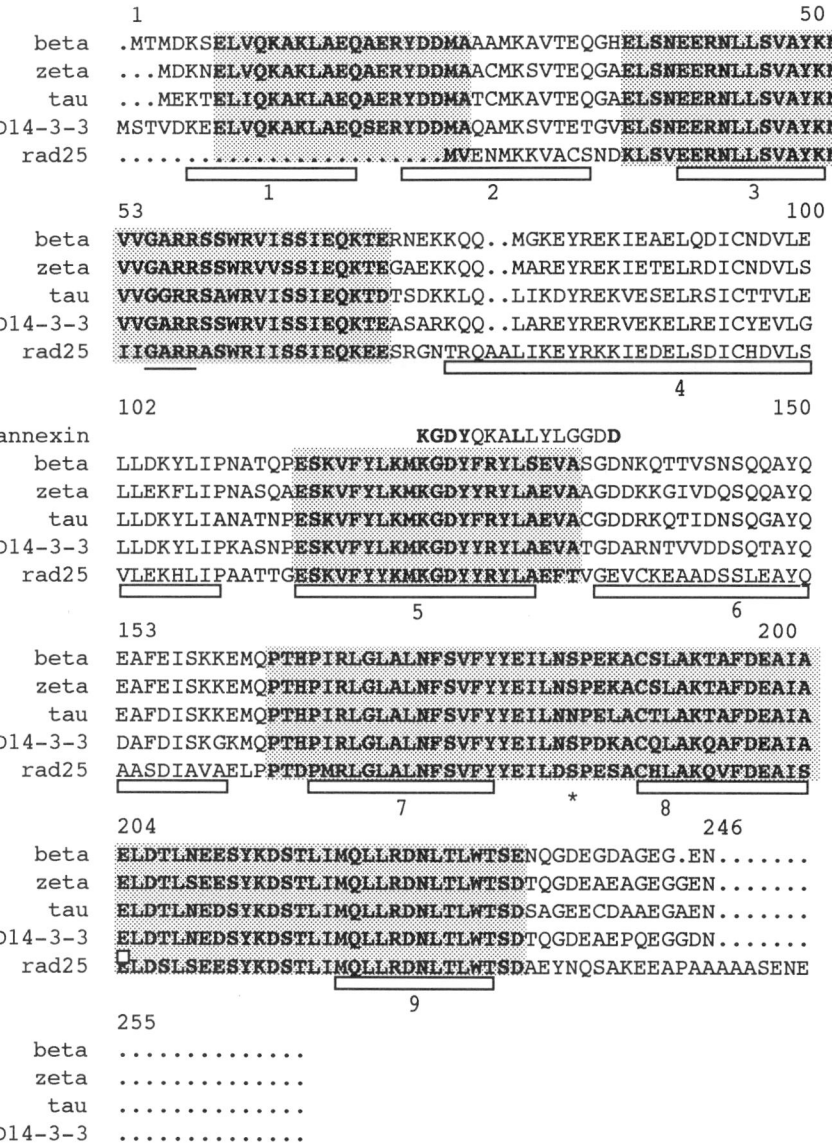

FIG. 1. Protein sequence alignment of 14-3-3 proteins. The human beta (β), zeta (ζ), and tau (τ) 14-3-3 isoforms are aligned with each other and 14-3-3 proteins from other organisms. Rad25 and D14-3-3 are isoforms from the yeast *Schizosaccharomyces pombe* and the fruit fly *Drosophila melanogaster*, respectively. Amino acid identity is present throughout the protein. Regions of highest conservation are shaded. The small region of the annexin family of proteins that shares homology with 14-3-3 protein is also shown in the alignment. The PKC pseudosubstrate region of 14-3-3 proteins is underlined. The amino acids that constitute each of the nine α-helices of a 14-3-3 monomer are indicated by a rectangle and the helix number is shown underneath the sequence. The site of *in vivo* phosphorylation of the β and ζ isoforms is indicated by an asterisk.

It has been shown that 14-3-3 proteins exist as dimers (Boston et al., 1982a). It is believed that 14-3-3 proteins homodimerize as well as heterodimerize with other isoforms (Jones et al., 1995a). Dimerization requires the amino terminus, which is not highly conserved among family members (Jones et al., 1995b; Liu et al., 1995; Xiao et al., 1995; Luo et al., 1995).

Recently, the crystal structures of the τ and ζ isoforms of 14-3-3 were reported (Xiao et al., 1995; Liu et al., 1995). As predicted by the amino acid sequence, the structure of the 14-3-3 dimer is predominantly α-helical. Each monomer consists of nine α-helices. The three amino-terminal helices of each monomer are involved in dimerization. This supports earlier work indicating that the amino terminus contains the dimerization region (Jones et al., 1995b). The dimerization interactions are primarily hydrophobic, with some electrostatic interactions. Analysis of the conservation of amino-terminal residues suggested that the low level of conservation may limit heterodimerization possibilities. However, the 3-dimensional analyses indicate that the residues involved in the dimerization of 14-3-3 proteins are in fact highly conserved (Liu et al., 1995). This suggests that 14-3-3 heterodimerization may be promiscuous in that all isoforms within a single cell may heterodimerize. Heterodimerization, however, may also be regulated by differential subcellular localization of 14-3-3 isoforms.

14-3-3 proteins are highly conserved at the amino acid level. Interestingly, the conserved residues of 14-3-3 proteins line the surface of a groove or channel of the ζ and τ 14-3-3 structures. Residues that are not conserved are located on the outside of this region (Xiao et al., 1995; Liu et al., 1995). This implicates the channel of 14-3-3 dimers as playing a key structural role required for 14-3-3 function.

Many targets of 14-3-3 are protein kinases (Aitken, 1995; Morrison, 1994). Interestingly, the Raf, PKC, and Bcr protein kinases are believed to interact with 14-3-3 proteins through basic zinc finger or cysteine-rich rich regions (Fu et al., 1994; Freed et al., 1994; Michaud et al., 1995; Reuther et al., 1994; Robinson et al., 1994). Side chains of the conserved residues of the 14-3-3 τ channel create an acidic region (Xiao et al., 1995). Thus, it is possible the basic residues in these protein kinases may interact with the conserved acidic region of the 14-3-3 channel. Further evidence that PKC may bind to this region of 14-3-3 is the presence within this channel of a region homologous to the annexins, a protein family with members that interact with PKC (Fig. 1) (Xiao et al., 1995; Mochly-Rosen et al., 1991a). Mutagenesis of this annexin homology, however, did not affect 14-3-3-mediated PKC regulation (Jones et al., 1995b).

The 14-3-3 carboxyl terminus, while not highly conserved at the amino acid level, contains many acidic residues. This acidic tail was found to interact with basic residues within the groove of the ζ crystal structure (Liu et al., 1995). It was proposed that this may provide a protein stability or regulatory role. In addition, a cluster of basic residues within the groove may be required for the phosphorylation-dependent interaction that is observed with Raf, Bcr, and tryptophan hydroxylases (Michaud et al., 1995; Furukawa et al., 1993). These interactions are described in the following sections.

The crystallographic studies have suggested that a 14-3-3 monomer may bind a protein helix and that helices of two cellular proteins may interact with a 14-3-3 dimer simultaneously (Liu et al., 1995). 14-3-3 proteins may be able to function as bridges that allow intracellular signaling molecules to interact. In this manner 14-3-3 proteins may play a role in coordinating cellular signaling events.

III. Regulation of Cellular Processes by 14-3-3 Proteins

A. Catecholamine Biosynthesis

Tryptophan and tyrosine hydroxylases catalyze the rate-limiting step in the biosynthesis of serotonin and noradrenaline, respectively (Cooper et al., 1986). Work by Yamauchi and Fujisama in the late 1970s and early 1980s indicated that activation of these hydroxylases required both a calcium–calmodulin-dependent protein kinase and an unidentified activator protein (Yamauchi et al., 1981). This protein was shown to have an apparent molecular weight of 35 kDa and to exist as a dimer. This activator protein was identified as brain 14-3-3 (Ichimura et al., 1987).

The mechanism of activation of the hydroxylases by calcium–calmodulin-dependent protein kinase and 14-3-3 protein has remained elusive. However, a potential mechanism has been suggested (Ichimura et al., 1988). The amino terminus of tryptophan and tyrosine hydroxylases serves as a regulatory domain for the catalytic carboxyl terminus. Phosphorylation of the hydroxylase would lead to 14-3-3 binding. Binding of the 14-3-3 acidic carboxyl terminus to basic residues of the hydroxylase regulatory region may result in a conformational change that activates the enzyme. Polyanions have been shown to activate these hydroxylases in a similar mechanism (Isobe et al., 1991). All 14-3-3 isoforms in the brain have been shown to activate tryptophan hydroxylase.

The phosphorylation-dependent interaction of 14-3-3 and tyrosine and tryptophan hydroxylases has been demonstrated *in vitro* (Furukawa *et al.*, 1993). Purified 14-3-3 proteins were shown to bind to tryptophan hydroxylase only after the enzyme was phosphorylated by calcium–calmodulin-dependent protein kinase. The same report showed that 14-3-3 did not affect the rate of tryptophan hydroxylase phosphorylation. Thus, the activating effect of 14-3-3 is likely due to its interaction with the hydroxylase after the enzyme is phosphorylated. It was reported, however, that 14-3-3 proteins could not activate a recombinant tyrosine hydroxylase (Sutherland *et al.*, 1993). Additional work is needed to understand the role of 14-3-3 proteins in the synthesis of serotonin and catecholamines.

While it appears that 14-3-3 proteins play a role in the *in vitro* activation of tryptophan and tyrosine hydroxylases, the presence of 14-3-3 in tissues outside of the nervous system indicates that these proteins have additional functions within the cell.

B. Exocytosis

14-3-3 proteins have recently been implicated in the regulation of exocytosis and secretion (Morgan and Burgoyne, 1992a,b; Wu *et al.*, 1992; Roth *et al.*, 1993). Permeabilized cells are used to study exocytosis. Cell permeabilization leads to the leakage of cytosolic proteins out of the cell. Adding back cytosolic proteins to the permeabilized cells and measuring exocytosis provides a mechanism by which proteins involved in exocytosis can be identified (Baker and Knight, 1981; Sarafian *et al.*, 1987). By using such a technique, Morgan and Burgoyne (1992a) described the involvement of two proteins, Exo1 and Exo2, in the process of exocytosis. These proteins were found to stimulate the calcium-dependent exocytosis from permeabilized adrenal chromaffin cells. Peptide sequencing of Exo1 indicated that it was a member of the 14-3-3 family. Antibodies raised against 14-3-3 proteins have been shown to inhibit the reactivation of exocytosis in permeabilized cells by exogenous cytosolic protein (Wu *et al.*, 1992). The mechanism of 14-3-3–mediated activation of exocytosis is unknown but may involve PKC. PKC and 14-3-3 can reactivate exocytosis in permeabilized cells. Addition of both of these proteins back to the cells has a synergistic effect on exocytosis (Morgan and Burgoyne, 1992a). The 14-3-3 effect on exocytosis requires Mg-ATP, suggesting the involvement of protein phosphorylation or perhaps a requirement for energy (Morgan and Burgoyne, 1992b). The authors concluded, however, that PKC phosphorylation of 14-3-3 proteins is not involved because PKC could not phosphorylate purified 14-3-3 *in vitro* (Morgan and Burgoyne, 1992b). However, *in vitro* phosphorylation conditions may not accurately emulate the PKC–14-3-3 environment within the cell, and therefore analysis of *in vivo* phosphorylation of 14-3-3 by PKC should be utilized to examine this issue. In contrast, another report has shown PKC phosphorylation of 14-3-3 *in vitro* (Jones *et al.*, 1995b). The existence of multiple isoforms of 14-3-3 and PKC will likely complicate analysis of this issue.

Annexin II is another protein capable of activating exocytosis in permeabilized chromaffin cells (Ali *et al.*, 1989). The annexins are a protein family shown to be involved in the fusion of membrane vesicles (Gruenberg and Emans, 1993). It is believed that annexin II is involved in the fusion of secretory vesicles to the plasma membrane. Interestingly, comparison of the primary sequences of annexin II with 14-3-3 proteins reveals a homology over a 16-amino acid region (Fig. 1) (Aitken *et al.*, 1990, 1992). A synthetic peptide corresponding to these amino acids inhibited Exo1-stimulated exocytosis (Roth *et al.*, 1993). This 16-amino acid region has been implicated in an interaction with PKC (Robinson *et al.*, 1994). The relationship between this region, PKC, and exocytosis is unknown. The potential function of the annexin homology in 14-3-3 proteins is discussed in a later section. While the mechanism of 14-3-3-mediated reactivation of exocytosis remains unclear, it has been shown that 14-3-3 proteins bind to chromaffin granules within the cell and to phospholipid vesicles *in vitro* (Roth *et al.*, 1994). The binding to vesicles was dependent on the isoform analyzed and thus may reveal isoform-specific functions within the cell.

C. ADP RIBOSYLATION

Proteins of the 14-3-3 family have been shown to be required for the enzymatic activity of a bacterial toxin (Fu *et al.*, 1993). A virulence factor of *Pseudomonas aeruginosa* is a protein called exoenzyme S (ExoS). ExoS is an ADP ribosyltransferase that places ADP ribose groups on small GTP-binding proteins *in vitro*. These include H-Ras, Rap1A, Rab3, and Rab4 (Coburn *et al.*, 1989; Coburn and Gill, 1991). The *in vivo* targets of ExoS in mammalian cells remain unknown. ExoS ADP ribosylation of a small GTP-binding protein in cells is likely to be an important step in the pathogenesis of *P. aeruginosa*. It is thought that infection disrupts vesicle movement within the cell (Coburn and Gill, 1991). Interestingly, Rab3 and Rab4 are involved in this process (Bourne *et al.*, 1990). This modification does not appear to alter the ability of GTP-binding proteins to interact with or hydrolyze guanine nucleotides. Instead it may prevent small G-proteins from interacting with key cellular targets. ADP ribosylation of cellular proteins by bacterial toxins is not uncommon. GTP-binding proteins are ADP ribosylated by botulinum C3, pertussis, and cholera toxins (Aktories, 1994; Gierschik, 1992; Moss et al., 1994).

The activity of ExoS was shown to require the presence of a cellular protein that was termed FAS (factor activating ExoS) (Fu et al., 1993). FAS was subsequently identified as the ζ isoform of the 14-3-3 protein family. Considering the extremely high conservation among 14-3-3 isoforms, it is not surprising that other family members have subsequently been shown to stimulate the ADP ribosylation activity of ExoS (Reuther et al., 1994; Chen et al., 1994; Lu et al., 1994). While it is apparent that ExoS requires the presence of 14-3-3 proteins to perform its ADP ribosylation function, the mechanism involving 14-3-3 in this process is unknown. 14-3-3 proteins may interact with the enzyme and cause an activating conformational change. Alternatively, 14-3-3 proteins may bring the enzyme in proximity to its substrate. It is unclear if 14-3-3 proteins serve a similar function within mammalian cells. There are ADP ribosyltransferases in mammalian cells, and it is therefore possible that 14-3-3 proteins may be involved in the control of ADP ribosylation of cellular proteins by these endogenous enzymes (Maehama et al., 1991). Nevertheless, the identification of 14-3-3 proteins as cofactors for ExoS provides a sensitive biochemical assay to study the 14-3-3 protein family (Pallas et al., 1994; Fu et al., 1994).

D. CELL CYCLE

Members of the 14-3-3 family are not restricted to higher eukaryotic organisms. Both the budding yeast *Saccharomyces cerevisiae* and the fission yeast *Schizosaccharomyces pombe* have two genes encoding 14-3-3 proteins (van Heusden et al., 1992, 1995; Ford et al., 1994).

In *S. cerevisiae* the two 14-3-3 genes are *BMH1* and *BMH2*. When the *BMH1* gene was either deleted or placed on a high-copy plasmid, the yeast were viable but exhibited altered (slower) growth rates (van Heusden et al., 1992). Likewise, yeast with a disrupted *BMH2* gene were viable. In contrast, disruption of both *BMH1* and *BMH2* was lethal. Thus, these genes together provide an essential function in budding yeast. It is likely that the two yeast 14-3-3 proteins, which are nearly identical except for their amino termini, have similar functions within the cell. This is supported by the fact that expression of a plant 14-3-3 protein in the double knockout allowed viable strains to be isolated. However, extended growth studies revealed defects that implicate yeast 14-3-3 as playing an important role near the end of the cell cycle, at mitosis or cell–cell separation (van Heusden et al., 1995).

Further evidence for a function of 14-3-3 proteins at mitosis comes from studies in the fission yeast *S. pombe*. Ford et al. (1994) described the products of the *rad24* and *rad25* genes as 14-3-3 family members that were involved in the DNA damage checkpoint before mitosis. Single knockouts of *rad24* and *rad25* produced viable yeast that entered mitosis prematurely following DNA damage by ionizing radiation. Disruption of both *rad24* and *rad25* resulted in a lethal phenotype. Thus in both budding and fission yeast there are two 14-3-3 proteins that are required for cell growth and appear to play an important role prior to mitosis.

While 14-3-3 proteins appear to play a role in the yeast cell cycle, less is known about their role in the mammalian cell cycle. A direct connection between 14-3-3 proteins and the cell cycle machinery has been made with the finding of the interaction between these proteins and the Cdc25 phosphatase (Conklin et al., 1995). Cdc25 is a phosphothreonine–phosphotyrosine phosphatase that dephosphorylates the cyclin-dependent kinase Cdc2. This leads to activation of Cdc2 kinase activity and facilitates entry into mitosis (Coleman and Dunphy, 1994). In *Xenopus*, 14-3-3 proteins have been shown to increase Raf-dependent Cdc2 kinase activation (Fantl et al., 1994). It is unknown whether 14-3-3 proteins directly affect Cdc25 phosphatase activity in cells. Cdc25 phosphorylation by the Raf kinase, another 14-3-3 interacting protein, may activate the phosphatase (Galaktionov et al., 1995). Thus, it is possible that 14-3-3 activation of Cdc2 in *Xenopus* may be mediated in part by Cdc25 activation via the Raf kinase. Interestingly, the defect observed in yeast when 14-3-3 genes are deleted is a defect at mitosis, and yeast Cdc25 is implicated at this cell cycle checkpoint (Ford et al., 1994; van Heusden et al., 1995).

A report by Pallas et al. (1994) indicates that 14-3-3 proteins associate with the middle T antigen (MT) of murine polyomavirus. MT associates with cellular proteins involved in signal transduction. This association with cellular proteins is believed to be important for the cellular transforming activity of MT. The biological significance of the association of 14-3-3 proteins with MT is unclear, but this interaction may be required for MT-mediated transformation. Interestingly, a 14-3-3 protein has been found to be downregulated in a number of human mammary carcinoma cell lines (Prasad et al., 1992). It is possible that 14-3-3 proteins participate in the regulation of cell cycle progression and that loss or inhibition of this activity may be a step in cellular transformation. Association with the MT viral protein may provide a mechanism for inhibiting a function of 14-3-3 proteins.

E. OTHER ACTIVITIES

To date only a single enzymatic function has been attributed to 14-3-3 proteins. Zupan *et al.* (1992) have described a phospholipase A_2 (PLA_2) activity associated with the ζ 14-3-3 isoform. The researchers showed that this 14-3-3 protein was able to cleave arachidonic acid from choline and ethanolamine phospholipids. Arachidonic acid has been shown to stimulate exocytosis. This finding is intriguing in light of the role of Exo1 (14-3-3) in exocytosis. Two years after the association between 14-3-3 proteins and PLA_2 was demonstrated, a report indicated that 14-3-3 proteins do not have PLA_2 activity (Robinson *et al.*, 1994).

The 14-3-3 ζ isoform was recently found to associate with the platelet glycoprotein Ib-IX complex (Du *et al.*, 1994). Glycoprotein Ib-IX is the receptor for von Willebrand factor (Sakariassen *et al.*, 1979). Binding of this ligand leads to platelet activation. While poorly understood, an early signaling event in the activation of platelets is PLA_2 activation. It was suggested that direct activation of PLA_2 (14-3-3 ζ) may occur upon ligand binding to glycoprotein Ib-IX (Du *et al.*, 1994). However, more convincing data are needed on any 14-3-3-associated PLA_2 activity to speculate on the significance of 14-3-3 protein binding to glycoprotein Ib-IX.

IV. 14-3-3 PROTEINS AND PROTEIN KINASES

Several 14-3-3 isoforms have been shown to interact with the protein kinases PKC, cRaf, cBcr, and Bcr-Abl (Aitken, 1995; Burbelo and Hall, 1995; Morrison, 1994). This section discusses the current understanding of the interactions of 14-3-3 proteins with these kinases.

A. PROTEIN KINASE C

Some of the most extensive studies of 14-3-3 proteins have focused on their role in the regulation of PKC. Toker *et al.* (1992) showed that PKC inhibitors (named KCIP) isolated from sheep brain were members of the 14-3-3 family. Initial studies indicated that 14-3-3 proteins inhibited PKC activity *in vitro*. This inhibition was not due to competition with required factors of the kinase reaction, such as calcium, ATP, substrate, or phosphatidylserine (Toker *et al.*, 1990). However, the inhibition could partly be prevented by increased concentrations of diacylglycerol or phorbol esters (Toker *et al.*, 1990; Robinson *et al.*, 1994). In addition, the inhibition was not due to an increase in phosphatase activity (Toker *et al.*, 1990). It is likely that the inhibition by 14-3-3 proteins involves the regulatory domain of PKC since the PKC catalytic domain alone was not inhibited by 14-3-3 proteins (Toker *et al.*, 1992).

Several mechanisms of 14-3-3 inhibition of PKC have been proposed. A stretch of amino acids in 14-3-3 exhibit homology to the pseudosubstrate region of the regulatory domain of PKC (Fig. 1) (Toker *et al.*, 1992; Aitken *et al.*, 1992). These sequences contain a consensus PKC phosphorylation site where the phosphorylatable residue is replaced by an alanine and thus cannot be phosphorylated (Parker *et al.*, 1989). Interaction with the pseudosubstrate sequence is believed to lock the catalytic domain of the kinase in an inactive state. The pseudosubstrate mechanism of inhibition is believed to be employed by many protein kinases (Kemp *et al.*, 1994). Thus, binding of the pseudosubstrate homologous region of 14-3-3 proteins to PKC may lead to PKC inhibition. Another potential mechanism of PKC inhibition by 14-3-3 proteins involves the previously described homology with the annexins and the potential involvement of this region in regulating the subcellular localization of PKC (Aitken *et al.*, 1990; Roth *et al.*, 1993; Mochly-Rosen, 1995). It is known that PKC activation leads to its translocation to the cytoskeleton (Mochly-Rosen, 1995). Proteins involved in the recruitment of PKC to cytoskeletal structures have been named *r*eceptors for *a*ctivated *C k*inase (RACK). Members of the annexin family have been shown to serve as RACK (Mochly-Rosen *et al.*, 1991a). Moreover, a peptide corresponding to the annexin homology within 14-3-3 proteins has been shown to prevent PKC binding to RACK (Mochly-Rosen *et al.*, 1991b). Therefore, 14-3-3 family members may be involved in the regulation of the translocation of PKC to the cytoskeleton upon its activation. Interestingly, site-directed mutagenesis of both the pseudosubstrate and annexin homologies in 14-3-3 proteins did not affect the ability of the mutant 14-3-3 proteins to inhibit PKC (Jones *et al.*, 1995b). However, a consensus peptide containing the annexin homology may be able to block a 14-3-3–PKC interaction (Xiao *et al.*, 1995). A more complete mutagenesis of this region will be required to test its involvement in potential 14-3-3–PKC interactions.

Biochemical analysis of the mechanism of PKC inhibition by 14-3-3 proteins has revealed a potential role for the cysteine-rich (C1) regulatory domain of PKC (Robinson *et al.*, 1994). This region has been shown to bind to the second messenger diacylglycerol and phorbol 12-myristate 13-acetate (PMA). Increasing concentrations of both diacylglycerol and PMA partially prevented the inhibition by 14-3-3 proteins. It is possible that 14-3-3 may interact with the C1 domain of PKC and thus interfere with the binding of this cysteine-rich regulatory region of PKC to diacylglycerol.

The nature of the relationship between PKC and 14-3-3 proteins is controversial. 14-3-3 proteins have been shown to activate as well as inhibit PKC activity (Isobe *et al.*, 1992; Tanji *et al.*, 1994). The ζ, β, and ϵ isoforms have been shown to activate PKC *in vitro* (Isobe *et al.*, 1992). This activation required the presence of the PKC cofactors calcium and phosphatidylserine. Another report indicated that calcium and phosphatidylserine were not required for 14-3-3-induced activation of PKC (Tanji *et al.*, 1994). Activation by 14-3-3 was specific to PKC as these proteins did not activate other kinases tested. While activation of tyrosine hydroxylase appears to require the correct secondary structure of 14-3-3 proteins, activation of PKC does not. Thus, the authors suggested that the interaction between 14-3-3 proteins and these two enzymes may be different. Recent studies with recombinant 14-3-3 proteins showed that they inhibit PKC similarly to purified brain 14-3-3 (Robinson *et al.*, 1994).

It has been proposed that activation of PKC by 14-3-3 proteins is mediated by 14-3-3 binding to histone, the substrate used in the *in vitro* PKC kinase reactions (Chen and Wagner, 1994). Binding of 14-3-3 to phosphorylated histone has been shown to protect the histone from the action of protein phosphatases. 14-3-3 activation of PKC did not occur when substrates other than histone were used in the reaction. Thus, the *in vitro* activation of PKC is specific to the substrate used in the assay and may not accurately represent a function of 14-3-3 *in vivo*.

The varying sources of 14-3-3 proteins utilized as well as the fact that there are multiple 14-3-3 isoforms could represent problems in attempting to generalize a role for 14-3-3 proteins in PKC regulation. Since PKC activation in cells is complex (i.e., a requirement for second messenger binding and translocation to the cytoskeleton), *in vitro* studies may not accurately reveal the significance of the relationship between PKC and 14-3-3 (Mochly-Rosen, 1995). Furthermore, the high concentrations (up to 1% in some cell types) of 14-3-3 proteins in cells requires that a 14-3-3-mediated regulation of PKC must itself be highly regulated (Boston *et al.*, 1982b). The phosphorylated forms of the β

and ζ isoforms of 14-3-3 were shown to have a twofold greater PKC inhibitory activity than their unphosphorylated counterparts (Aitken et al., 1995). Different subcellular localization of PKC and 14-3-3 proteins may be another mode for the regulation of PKC activity by 14-3-3 proteins. It is important to note that presently there is no evidence that 14-3-3 proteins associate with or regulate PKC in cells.

B. Raf

A great interest in 14-3-3 proteins has recently been stimulated by the discovery that 14-3-3 proteins interact with the protein kinase Raf (Freed et al., 1994; Fu et al., 1994; Fantl et al., 1994; Li et al., 1995). In cells, Raf is activated by an increase in the levels of GTP-bound Ras. Binding of receptor tyrosine kinases to their corresponding ligands leads to an increase in the GTP-bound form of Ras. Ras–GTP then interacts with Raf, causing the translocation of Raf to the plasma membrane. The membrane-associated Raf is activated by an unknown mechanism, which may involve a protein kinase (Daum et al., 1994; Marais et al., 1995).

Raf has been shown to interact with several isoforms of the 14-3-3 family. This interaction has been shown through the use of several different experimental techniques, including both genetic and molecular biological approaches. Freed et al. (1994) identified the β and ζ 14-3-3 isoforms as Raf-interacting proteins using the yeast two-hybrid screen. The 14-3-3 proteins interacted with the cysteine-rich conserved region 1 (CR1) region and the serine-rich CR2 region of Raf, as well as the carboxyl-terminal kinase domain. A second group showed that proteins associated with Raf upon immunoprecipitation were 14-3-3 family members (Fu et al., 1994). This study also presented data supporting the association of 14-3-3 proteins with the CR1, CR2, and kinase domain of Raf. Several other groups have also shown that Raf interacts with 14-3-3 proteins (Fantl et al., 1994; Irie et al., 1994; Li et al., 1995; Yamamori et al., 1995; Luo et al., 1995). Luo et al. (1995) showed that the carboxyl-terminal 100 amino acids of 14-3-3 ζ are involved in the interaction with Raf.

Since Raf plays a critical role in signal transduction pathways elicited by receptor tyrosine kinases, the relationship between the Raf–14-3-3 interaction and Raf kinase activity is important. Catalytically active and inactive Raf kinases interact with 14-3-3 proteins (Fu et al., 1994; Freed et al., 1994; Michaud et al., 1995). The 14-3-3 proteins have been found to associate with Raf prior to its activation at the plasma membrane (Freed et al., 1994). Li et al. (1995) suggest that in fibroblasts 14-3-3 proteins do not associate with Raf that has been activated by mitogenic stimulation. The Raf–14-3-3 interaction, however, has been shown *in vitro* to be phosphorylation dependent (Michaud *et al.*, 1995). Kinase-inactive Raf expressed in cells is phosphorylated on several residues by other kinases. In fact, mutation of a serine residue to alanine abolishes Raf association with 14-3-3 *in vivo* (Michaud *et al.*, 1995). While this phosphorylatable residue in Raf may be critical for direct interaction with 14-3-3, it is possible that this mutation could have produced a structural change in Raf that disrupted the 14-3-3 interaction.

Irie *et al.* (1994) used a screen in yeast to identify proteins that, when overexpressed, could activate Raf. The yeast 14-3-3 protein Bmh1 was identified by this technique. Thus, 14-3-3 was implicated in the *in vivo* activation of the Raf kinase. Interestingly Bmh1 was found to be required for Raf activation by yeast Ras protein. Similar results were obtained in yeast by Freed *et al.* (1994). Raf can also be activated in *Xenopus* oocytes by coexpression with 14-3-3 (Fantl *et al.*, 1994). One report has shown Raf activation by 14-3-3 proteins in *in vitro* kinase assays (Irie *et al.*, 1994). However, many other reports have indicated 14-3-3 proteins could not activate Raf *in vitro* (Fu *et al.*, 1994; Li *et al.*, 1995; Michaud *et al.*, 1995; Luo *et al.*, 1995). These contradicting results may be due to the different preparations of bacterial 14-3-3 utilized, the different assays used to measure Raf kinase activity, or the requirement for posttranslational modification of 14-3-3 for activation of Raf. In addition, it has been shown that 14-3-3 proteins could activate REKS, a complex of B-Raf and 14-3-3 proteins, in a cell-free system (Shimizu *et al.*, 1994; Yamamori *et al.*, 1995).

While an increase in Ras–GTP levels leads to activation of Raf *in vivo*, controversy exists regarding the ability of 14-3-3 proteins to activate Raf in cells. Overexpression of 14-3-3 can activate Raf and Raf-dependent processes (Freed *et al.*, 1994; Irie *et al.*, 1994; Fantl *et al.*, 1994; Li *et al.*, 1995). 14-3-3 proteins do not activate Raf *in vitro* and can interact with inactive Raf (Fu *et al.*, 1994; Freed *et al.*, 1994; Li *et al.*, 1995; Michaud *et al.*, 1995). Therefore, 14-3-3 proteins do not appear to directly activate Raf. In addition, point mutants of Raf that do not interact with 14-3-3 are kinase active in cells (Michaud *et al.*, 1995). However, the finding that Raf contains multiple 14-3-3 binding sites makes these results difficult to interpret.

An interesting report demonstrated that protein phosphatases in membrane fractions can decrease Raf kinase activity, presumably through dephosphorylation of the kinase (Dent *et al.*, 1995). In this study, 14-3-3 protein was shown to protect Raf from inactivation by phosphatases, presumably through a direct phosphorylation-dependent interaction of 14-3-3 with Raf. Thus, specific phosphate groups on Raf would be inaccessible to phosphatases because of the bound 14-3-3 (Fig. 2). These observations may explain how overexpression of 14-3-3 activates Raf in cells. In fact, it is possible that 14-3-3 overexpression does not activate Raf, but rather inhibits raf inactivation, resulting in a measurable increase in Raf activity.

It is believed that 14-3-3 interacts with inactive cytosolic Raf and translocates to the plasma membrane with Raf upon activation of Ras (Freed et al., 1994). At the membrane, Raf may be activated by phosphorylation (Fig. 2). It is unclear whether 14-3-3 proteins remain bound to Raf after Raf activation following mitogenic stimulation. While a 14-3-3 function may be to protect phosphorylated Raf from phosphatase activity (Fig. 2), these proteins may not interact with this region of Raf before phosphorylation, since they could hinder the kinase from phosphorylating Raf (Freed et al., 1994; Fu et al., 1994; Li et al., 1995). It is therefore possible that 14-3-3 interacts with inactive cytosolic Raf at a given region of Raf. After Raf activation (presumably by phosphorylation), 14-3-3 bound to Raf may shift to protect the newly phosphorylated residues from phosphatase activity. Alternatively, 14-3-3 proteins may bind to multiple regions of Raf simultaneously.

Several other mechanisms of 14-3-3-mediated activation of Raf are possible. Following Raf activation, 14-3-3 proteins may simply function to stabilize the conformation of the active kinase (Fig. 2). Since 14-3-3 proteins dimerize and are known to interact with other proteins, they may facilitate an interaction with a protein coactivator of Raf (Fig. 2). Complexes of Raf and Bcr, another 14-3-3-interacting protein, have been observed in mammalian cells overexpressing Bcr (Braselmann and McCormick, 1995; G. W. Reuther and A. M. Pendergast, unpublished data). The Cdc25 phosphatase interacts with both 14-3-3 proteins and Raf (Galaktionev et al., 1995; Conklin et al., 1995). It is unknown whether 14-3-3 proteins mediate Raf–Bcr or Raf–Cdc25 complexes in cells. Additionally, Raf is believed to translocate to the cytoskeletal compartment after activation, similar to PKC (Stokoe et al., 1994). 14-3-3 proteins may facilitate Raf translocation and/or interaction with cytoskeletal components. The annexin homology region, which is implicated in the binding of 14-3-3 proteins to PKC, may facilitate such cytoskeletal interactions (see Section IV,A). Both PKC and Raf have similar cysteine-rich motifs* that are implicated in 14-3-3 interaction (Fig. 3). It will be interesting to examine whether a peptide consisting of the 14-3-3 annexin homology region affects Raf activa-

---

*See Note added in proof.

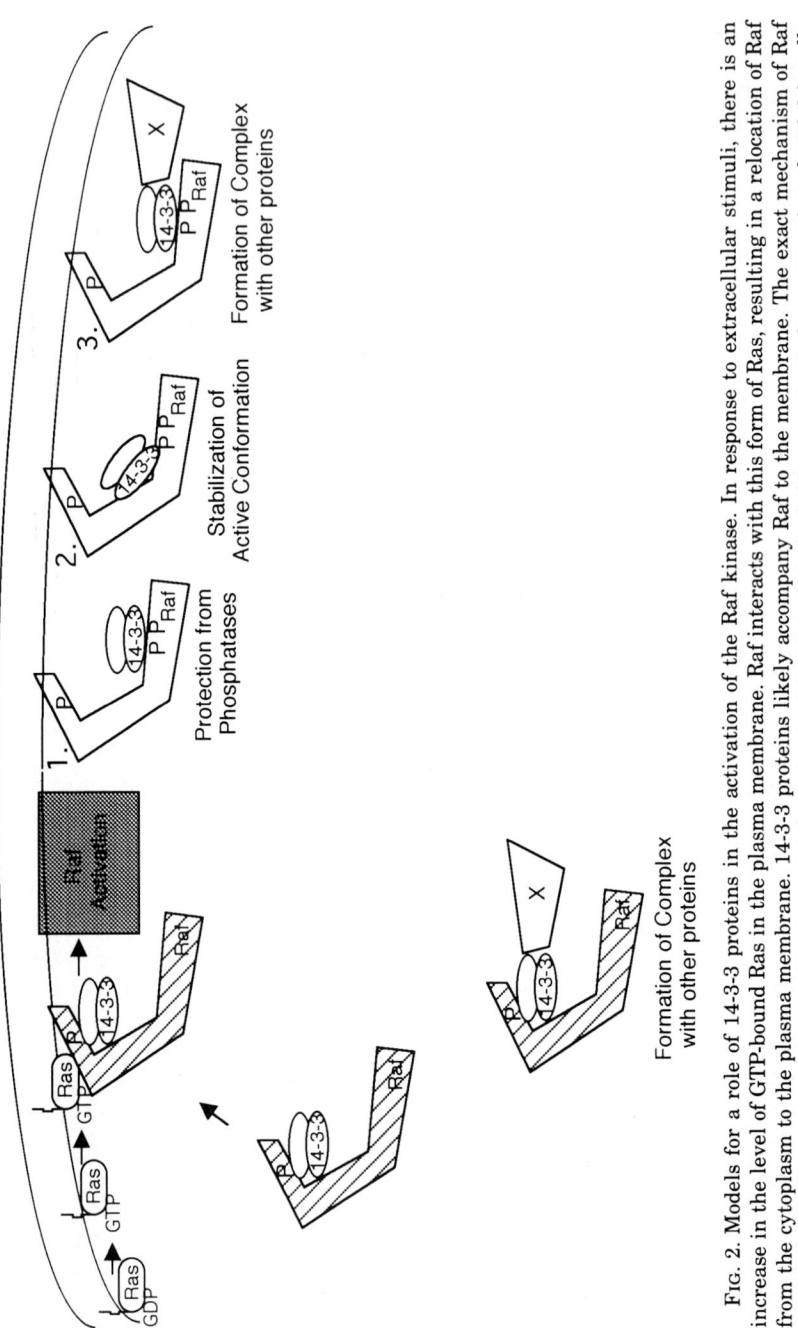

FIG. 2. Models for a role of 14-3-3 proteins in the activation of the Raf kinase. In response to extracellular stimuli, there is an increase in the level of GTP-bound Ras in the plasma membrane. Raf interacts with this form of Ras, resulting in a relocation of Raf from the cytoplasm to the plasma membrane. 14-3-3 proteins likely accompany Raf to the membrane. The exact mechanism of Raf activation at the plasma membrane is not completely understood (indicated by shaded box; see text). Overexpression of 14-3-3 in cells has been shown to activate Raf. After this activation, 14-3-3 proteins may bind to phosphorylated active Raf at specific sites and protect Raf from dephosphorylation by protein phosphatases (1). These proteins may also help stabilize the active conformation of Raf (2). In addition, 14-3-3 proteins may facilitate Raf complex formation with other Raf coactivators (3).

CYSTEINE-RICH

```
Bcr          388C h-KrHrH CpvvV
Raf          168C gyKfHeH CstkV
Pkc              CcfvvHkrChefV
consensus       C   k H hC    V
```


SERINE-RICH

```
Bcr          298RsqstS eqqkrlTwprRsyS PrsFedcgggyT PdcS SnenLTss
Raf          215RrmreS vsrmpvSsqhRysT PhaF-tfnts-S P--S SegsLSqr
consensus       R    S       T   R SP  F          TP   SS   LT
                     S                                  S       S
```

```
Bcr          342eedfS SgqssrV SpS -PttyRMfrDksR SpSqnsqqS fdSSSP
Raf          255qrstS TpnvhmV StT lPvdsRMieDaiR ShSesaspS alSSSP
consensus       SS      VS S      P    RM  D  RS S        S SSSP
                T       T
```

FIG. 3. Amino acid alignment of cysteine- and serine-rich regions of Raf, Bcr, and PKC sequences implicated in 14-3-3 interactions. The serine-rich alignment of Bcr and Raf shows 28% identity and 34% similarity. The Raf and PKC cysteine-rich homologies extend beyond the sequences shown. Analysis of these sequences may provide a clue to the primary and/or secondary structures that are targeted by 14-3-3 proteins.

tion and translocation of Raf to the cytoskeleton. Finally, Raf associated with truncated mutants of 14-3-3 in cells is not activated, unlike Raf bound to full-length 14-3-3 proteins (Luo et al., 1995). Therefore, functional 14-3-3 protein may participate in Raf activation *in vivo*.

C. BCR AND BCR-ABL

The *bcr-abl* oncogene is formed by a chromosomal translocation that fuses the *bcr* gene upstream of the c-*abl* protooncogene. Bcr-Abl is an activated tyrosine kinase that is believed to play a causative role in chronic myelogenous leukemia and a subset of acute lymphocytic leukemias (Kurzrock et al., 1988). Bcr is a multidomain protein that encodes a serine kinase activity at the amino-terminus, a region of homology with guanine nucleotide exchange factors in the central portion of the molecule, and GTPase-activating activity in the carboxy terminus (Maru and Witte, 1991; Ron et al., 1991; Diekmann et al., 1991; Ridley et al., 1993). It has been shown that Bcr sequences in Bcr-Abl are required for the full cellular transforming activity of the chimeric Bcr-Abl protein (Muller et al., 1991; Pendergast et al., 1991, 1993; McWhirter et al., 1993; Reuther et al., 1994). We showed that the τ (T-cell) isoform of 14-3-3 associates with both the Bcr serine kinase and the Bcr-Abl tyrosine kinase (Reuther et al., 1994). Sequences of Bcr implicated in 14-3-3 binding are rich in serine and cysteine residues. Interestingly, the serine-rich region of Bcr has 34% homology with a region of Raf that binds 14-3-3 proteins (Fig. 3). Also, homology exists between the cysteine-rich regions of Bcr and Raf. Point mutations of Raf that prevent 14-3-3 binding are within these two homologous regions. Like Raf, Bcr interacts with 14-3-3 in a phosphorylation-dependent manner (Michaud et al., 1995). Bcr has also been shown to associate with the β isoform of 14-3-3 proteins (Braselmann and McCormick, 1995).

Bcr and Bcr-Abl can phosphorylate both 14-3-3 τ and ζ *in vitro* (Reuther et al., 1994). However, the τ isoform was more heavily phosphorylated in these experiments, suggesting that these two highly identical isoforms were recognized differently by these kinases and/or contain different sites of phosphorylation for these kinases. The τ isoform was the only 14-3-3 protein found in a screen for Bcr-interacting proteins. Bcr appears to be the principle kinase that phosphorylates the τ 14-3-3 isoform expressed in fibroblasts. The significance of the phosphorylation of 14-3-3 proteins by the Bcr kinase has yet to be determined.

The role of 14-3-3 proteins in signaling by Bcr-Abl or Bcr is unknown. The primary function of Bcr has been shown to be in the regulation of the oxidative burst of activated neutrophils (Voncken et al., 1995). The oxidative burst is created by a multicomponent system that includes the small guanine nucleotide-binding protein Rac (Diekmann et al., 1994). Bcr has been shown to exhibit GTPase-activating activity toward Rac (Diekmann et al., 1991; Ridley et al., 1993). 14-3-3 proteins may play a key role in the formation of a proper protein complex that is required for Bcr-mediated regulation of the oxidative burst in neutrophils.

V. Roles of 14-3-3 Proteins in Signal Transduction Pathways

It is clear that 14-3-3 proteins physically interact with a number of cellular proteins (Table I). These interactions often appear to be regulated by phosphorylation. Several roles for 14-3-3 proteins in signal transduction are suggested from the findings described earlier. Among these are roles as chaperones, molecular bridges, and effectors.

TABLE I
LIST OF PROTEINS AND KNOWN
ASSOCIATED 14-3-3 ISOFORMS[a]

Cellular protein	Associated isoform
Tryptophan hydroxylase	Multiple isoforms
Tyrosine hydroxylase	Multiple isoforms
Protein kinase C	Multiple isoforms
Raf	β, ζ, ε, others
Bcr	τ, ζ, β
Bcr-Abl	τ, ζ
Middle T antigen	ζ, ε
Glycoprotein Ib-IX	ζ
Cdc25	β, ε

[a] This table is not intended to show 14-3-3 isoform binding specificity with these proteins. It is very likely that multiple isoforms interact with these proteins. This table is based on both *in vivo* and *in vitro* data. Association with the hydroxylases is based on 14-3-3-mediated activation of the enzymes. A physical association between 14-3-3 proteins and PKC has not been established. This association is based on the inhibitory effects of 14-3-3 proteins toward PKC kinase activity.

A. CHAPERONES

Molecular chaperones, such as the heat shock proteins (hsp), are proteins that aid in the folding of newly synthesized polypeptide chains and translocation of proteins across membranes (Craig, 1993; Agard, 1993). Chaperones are usually released after proper protein folding. Sometimes, however, these proteins remain bound to functionally folded proteins within the cell. Raf and steroid receptors are examples of proteins that interact with hsp (Wartmann and Davis, 1994; Pratt and Welsh, 1994). Steroid receptor interaction with these proteins is required for proper function of the receptor (Pratt and Welsh, 1994). Casein kinase II and eIF-2α kinase activities are increased upon interaction of these kinases with hsp (Miyata and Yahara, 1992; Szyszka *et al.*, 1989). Thus, chaperones are important for the proper folding and function of many cellular proteins.

14-3-3 proteins may function as molecular chaperones. It is not known if 14-3-3 proteins aid in the proper folding of polypeptides. However, as discussed, 14-3-3 binding to enzymes has been shown to increase enzymatic activity. This may be due to a stabilization of the active kinase. Additional evidence that 14-3-3 proteins may function as chaperones comes from the work of Alam *et al.* (1994) showing that 14-3-3 proteins prevented protein aggregation *in vitro* and were required for protein import into mitochondria. These are functions associated with molecular chaperones. Therefore, it is possible that the 14-3-3 and hsp protein families are functionally related.

B. Molecular Bridges

Since 14-3-3 proteins function as dimers and have been shown to interact with many proteins, it is possible that a single 14-3-3 dimer may interact with two proteins simultaneously or allow for intramolecular interactions between different domains of the same protein. This idea is supported by the structural studies performed on 14-3-3 proteins (Xiao *et al.*, 1995; Liu *et al.*, 1995). As described earlier, 14-3-3 dimers contain a ligand-binding region the size of which may permit the binding of two α-helices. Therefore, 14-3-3 proteins may act as molecular bridges by associating two proteins into a single complex (Jones *et al.*, 1995a). Indeed, Raf and Bcr, both 14-3-3-interacting proteins, have been shown to be complexed in cells overexpressing Bcr (Braselmann and McCormick, 1995; G. W. Reuther and A. M. Pendergast, unpublished data). Raf and Cdc25 also interact in cells (Galaktionov *et al.*, 1995). The role of 14-3-3 proteins in these protein complexes is unknown. 14-3-3 proteins acting as molecular bridges could bring enzymes in proximity to their substrates and/or activators, allow for proper protein subcellular localization, and serve as junctions for cross-talk of signaling pathways. For example, the Bcr-Abl tyrosine kinase activates the Ras signaling pathway directly by interacting with the Grb2 adaptor protein and indirectly through the phosphorylation of several Grb2-binding proteins, including Shc (Pendergast *et al.*, 1993; Cortez *et al.*, 1995). Bcr-Abl may also be able to signal through this pathway via a 14-3-3-mediated connection to Raf.

C. Effectors

A topic that has not been thoroughly addressed is that of 14-3-3 protein phosphorylation. Members of this family have been shown to be phosphorylated in cells and *in vitro* by various protein kinases (Toker *et al.*, 1992; Reuther *et al.*, 1994; Jones *et al.*, 1995b). In fact, Aitken *et al.* (1995) have shown that the α and δ isoforms are actually the phosphorylated forms of the β and ζ isoforms. The phosphorylation site of these isoforms is serine 186 (corresponding to the human β isoform). This serine is in the sequence serine–proline–x–lysine, which is a consensus phosphorylation site for cyclin-dependent kinases. The β and ζ isoforms are the only human 14-3-3 proteins that contain this consensus site. It is not known if cyclin-dependent kinases phosphorylate these 14-3-3 proteins. Phosphorylation by these kinases could provide a link between 14-3-3 proteins and the cell cycle machinery. Phosphorylation of 14-3-3 proteins may affect Raf kinase and tyrosine and tryptophan hydroxylase activity. 14-3-3 phosphorylation has been suggested to affect 14-3-3-mediated inhibition of PKC (Aitken et al., 1995). Therefore, 14-3-3 proteins may act as signaling effectors via phosphorylation from both extracellular and cell cycle-dependent signaling pathways.

VI. Conclusions

While much has been learned about 14-3-3 proteins over the past decades, a definitive role for these proteins within the cell remains unknown. Regulation of processes such as hydroxylation, secretion, phosphorylation, and cell cycle checkpoints are suggested roles. The dimeric 14-3-3 proteins interact with numerous cellular proteins (Table I). In some cases these interactions depend on the phosphorylation of the protein targeted by 14-3-3 proteins. 14-3-3 proteins may act in the cell as protein chaperones, molecular bridges, or signaling effectors. The renewed interest in the function(s) of 14-3-3 proteins may produce answers to the many unresolved questions about the roles of these proteins in signal transduction and other cellular processes.

Acknowledgments

We thank Haian Fu for critically reading this manuscript and insightful comments. G. W. Reuther is supported by an Environmental Protection Agency Graduate Student Fellowship. Work in this laboratory was supported by National Institutes of Health grant CA61033 to A. M. Pendergast. A. M. Pendergast is a Whitehead Scholar and a Scholar of the Leukemia Society of America.

Note Added in Proof: While this manuscript was in preparation a significant finding on 14-3-3/protein interactions was reported (Muslin, A. J., Tanner, J. W., Allen, P. M., and Shaw, A. S. (1996) *Cell* **84**, 889–897). This work shows that 14-3-3 proteins bind their targets at sites of serine phosphorylation. The consensus 14-3-3 binding site is reported as arginine-serine-x-serine (phosphorylated)-x-proline. Although two 14-3-3 binding sites (serine-259 and serine-621) were identified on Raf, the authors conclude that the primary binding site is serine-621. Mutation of this serine renders the Raf kinase inactive and inhibits association with 14-3-3. Therefore, 14-3-3 interaction with Raf at serine-621 may be required for a functional Raf kinase.

REFERENCES

Agard, D. A. (1993). To fold or not to fold. . . . *Science* **260,** 1903–1904.
Aitken, A. (1995). 14-3-3 proteins on the map. *Trends Biochem Sci.* **20,** 95–97.
Aitken, A., Ellis, C. A., Harris, A., Sellers, L. A., and Toker, A. (1990). Kinase and neurotransmitters, *Nature (London)* **344,** 594.
Aitken, A., Collinge, D. B., van Heusden, B. P. H., Isobe, T., Rosebloom, P. H., Rosenfield, G., and Soll, J. (1992). 14-3-3 proteins: A highly conserved, widespread family of eukaryotic proteins. *Trends Biochem. Sci.* **17,** 498–501.
Aitken, A., Howell, S., Jones, D., Madrazo, J., and Patel, Y. (1995). 14-3-3 α and δ are the phosphorylated forms of Raf-activating 14-3-3 β and ζ *J. Biol. Chem.* **270,** 5706–5709.
Aktories, K. (1994). Clostridial ADP-ribosylating toxins: Effects on ATP and GTP-binding proteins. *Mol. Cell. Biochem.* **138,** 167–176.
Alam, R., Hachiya, N., Sakaguchi, M., Kawabata, S., Iwanaga, S., Kitajima, M., Mihara, K., and Omura, T. (1994). cDNA cloning and characterization of mitochondrial import stimulation factor (MSF) purified from rat liver cytosol. *J. Biochem.* **116,** 416–425.
Ali, S. M., Geisow, M. J., and Burgoyne, R. D. (1989). A role for calpactin in calcium-dependent exocytosis in adrenal chromaffin cells. *Nature (London)* **340,** 313–315.
Baker, P. F., and Knight, D. E. (1981). Calcium control of exocytosis and endocytosis in bovine adrenal medullary cells. *Philos. Trans. R. Soc. Lond. Ser. B* **296,** 83–103.
Boston, P. F., Jackson, P., Kynoch, P. A. M., and Thompson, R. J. (1982a). Purification, properties, and immunohistochemical localisation of human brain 14-3-3 protein. *J. Neurochem.* **38,** 1466–1474.
Boston, P. F., Jackson, P., and Thompson, R. J. (1982b). Human 14-3-3 protein: Radioimmunoassay, tissue distribution, and cerebrospinal fluid levels in patients with neurological disorders. *J. Neurochem.* **38,** 1475–1482.
Bourne, H. R., Sanders, D. A., and McCormick, F. (1990). The GTPase superfamily: A conserved switch for diverse cell functions. *Nature (London)* **348,** 125–132.
Braselmann, S., and McCormick, F. (1995). BCR and RAF form a complex *in vivo* via 14-3-3 proteins. *EMBO J.* **14,** 4839–4848.
Burbelo, P. D., and Hall, A. (1995). 14-3-3 proteins: Hot numbers in signal transduction. *Curr. Biol.* **5,** 95–96.
Chen, F., and Wagner, P. D. (1994). 14-3-3 proteins bind to histone and affect both histone phosphorylation and dephosphorylation. *FEBS Lett.* **347,** 128–132.
Chen, Z., Fu, H., Liu, D., Chang, P. F., Narasimhan, M., Ferl, R., Hasegawa, P. M., and Bressan, R. A. (1994). A NaCl-regulated plant gene encoding a brain protein homology that activates ADP ribosyltransferase and inhibits protein kinase C. *Plant J.* **6,** 729–740.
Coburn, J., and Gill, D. M. (1991). ADP-ribosylation of p21ras and related proteins by Pseudomonas aeruginosa exoenzyme S. *Infect. Immun.* **59,** 4259–4262.
Coburn, J., Wyatt, R. T., Iglewski, B. H., and Gill, D. M. (1989). Several GTP-binding proteins, including p21c-H-ras, are preferred substrates of Pseudomonas aeruginosa exoenzyme S. *J. Biol Chem.* **264,** 9004–9008.
Coleman, T. R., and Dunphy, W. G. (1994). Cdc2 regulatory factors. *Curr. Opin. Cell Biol.* **6,** 877–882.
Conklin, D. S., Galaktionov, K., and Beach, D. (1995). 14-3-3 proteins associate with cdc25 phosphatases. *Proc. Natl. Acad. Sci. U.S.A.* **92,** 7892–7896.

Cooper, J. R., Bloom, F. E., and Roth, R. H. (1986). "Biochemical Basis of Neuropharmacology," 5th ed., pp. 203–339. Oxford University Press, New York.
Cortez, D., Kadlec, L., and Pendergast, A. M. (1995). Structural and signaling requirements for BCR-ABL-mediated transformation and inhibition of apoptosis. *Mol. Cell. Biol.* **15**, 5531–5541.
Craig, E. A. (1993). Chaperones: Helpers along the pathways to protein folding. *Science* **260**, 1902–1903.
Daum, G., Eisenmann-Tappe, I., Fries, H. W., Troppmair, J., and Rapp, U. (1994). The ins and outs of Raf kinases. *Trends Biochem. Sci.* **19**, 474–480.
Dent, P., Jelinek, T., Morrison, D. K., Weber, M. J., and Sturgill, T. W. (1995). Reversal of Raf-1 activation by purified and membrane-associated protein phosphatases. *Science* **268**, 1902–1906.
Diekmann, D., Brill, S., Garrett, M. D., Totty, N., Hsuan, J., Monfries, C., Hall, C., Lim, L., and Hall, A. (1991). Bcr encodes a GTPase-activating protein for p21rac. *Nature (London)* **351**, 400–402.
Diekmann, D., Abo, A., Johnston, C., Segal., A. W., and Hall, A. (1994). Interaction of Rac with p67phox and regulation of phagocytic NADPH oxidase activity. *Science* **265**, 531–533.
Du, X., Harris, S. J., Tetaz, T. J., Ginsberg, M. H., and Berndt, M. C. (1994). Association of a phospholipase A2 (14-3-3 protein) with the platelet glycoprotein Ib-IX complex. *J. Biol. Chem.* **269**, 18287–18290.
Fantl, W. J., Muslin, A. J., Kikuchi, A., Martin, J. A., MacNicol, A. M., Gross, R. W., and Williams, L. T. (1994). Activation of Raf-1 by 14-3-3 proteins. *Nature (London)* **371**, 612–614.
Ford, J. C., Al-Khodairy, F., Fotou, E., Sheldrick, K. S., Griffiths, D. J. F., and Carr, A. M. (1994). 14-3-3 protein homologs required for the DNA damage checkpoint in fission yeast. *Science* **265**, 533–537.
Freed, E., Symons, M., Macdonald, S. G., McCormick, F., and Ruggieri, R. (1994). Binding of 14-3-3 proteins to the protein kinase Raf and effects on its activation. *Science* **265**, 1713–1715.
Fu, H., Coburn, J., and Collier, R. J. (1993). A eukaryotic host factor that activates exoenzyme S of *Pseudomonas aeruginosa* is a member of the 14-3-3 protein family. *Proc. Natl. Acad. Sci. U.S.A.* **90**, 2320–2324.
Fu, H., Xia, K., Pallas, D., Cui, C., Conroy, K., Narsimhan, R. P., Mamnon, H., Collier, R. J., and Roberts, T. M. (1994). Interaction of the protein kinase Raf-1 with 14-3-3 proteins. Science **266**, 126–129.
Furukawa, Y., Ikuta, N., Omata, S., Yamauchi, T., Isobe, T., and Ichimura, T. (1993). Demonstration of the phosphorylation-dependent interaction of tryptophan hydroxylase with the 14-3-3 proteins. *Biochim. Biophys. Act* **194**, 144–149.
Galaktionov, K., Jessus, C., and Beach, D. (1995). Raf1 interaction with Cdc25 phosphatase ties mitogenic signal transduction to cell cycle activation. *Genes Dev.* **9**, 1046–1058.
Gierschik, P. (1992). ADP-ribosylation of signal-transducing guanine nucleotide-binding proteins by pertussis toxin. *Curr. Top. Microbiol. Immunol.* **175**, 69–96.
Gruenberg, J., and Emans, N. (1993). Annexins in membrane traffic. *Trends Cell Biol.* **3**, 224–227.
Ichimura, T., Isobe, T., Okuyama, T., Yamauchi, T., and Fujisawa, H. (1987). Brain 14-3-3 protein is an activator protein that activates tryptophan 5-monooxygenase and tyrosine 3-monooxygenase in the presence of Ca^{2+}, calmodulin-dependent protein kinase II. *FEBS Lett.* **219**, 79–82.

Ichimura, T., Isobe, T., Okuyama, T., Takahashi, N., Araki, K., Kuwano, R., and Takahashi, Y. (1988). Molecular cloning of cDNA coding for brain-specific 14-3-3 protein, a protein kinase-dependent activator of tyrosine and tryptophan hydroxylases. *Proc. Natl. Acad. Sci. U.S.A.* **85**, 7084–7088.

Irie, K., Gotoh, Y., Yashar, B. M., Errede, B., Nishida, E., and Matsumoto, K. (1994). Stimulatory effects of yeast and mammalian 14-3-3 proteins on the Raf protein kinase. *Science* **265**, 1716–1719.

Isobe, T., Ichimura, T., Sunaya, T., Okuyama, T., Takahashi, N., Kuwano, R., and Takahashi, Y. (1991). Distinct forms of the protein kinase-dependent activator of tyrosine and tryptophan hydroxylases. *J. Mol. Biol.* **217**, 125–132.

Isobe, T., Hiyane, Y., Ichimura, T., Okuyama, T., Takahashi, N., Nakajo, S., and Nakaya, K. (1992). Activation of protein kinase C by the 14-3-3 proteins homologous with Exo1 protein that stimulates calcium-dependent exocytosis. *FEBS Lett.* **308**, 121–124.

Jones, D. H., Ley, S., and Aitken, A. (1995a). Isoforms of 14-3-3 protein can form homo- and heterodimers in vivo and in vitro: Implications for function as adapter proteins. *FEBS Lett.* **368**, 55–58.

Jones, D. H. A., Martin, H., Madrazo, J., Robinson, K. A., Nielsen, P., Roseboom, P. H., Patel, Y., Howell, S. A., and Aitken, A. (1995b). Expression and structural analysis of 14-3-3 proteins. *J. Mol. Biol.* **245**, 375–384.

Kemp, B. E., Parker, M. W., Hu, S., Tiganis, T., and House, C. (1994). Substrate and pseudosubstrate interactions with protein kinases: Determinants of specificity. *Trends Biochem. Sci.* **19**, 440–444.

Kurzrock, R., Gutterman, J., and Talpaz, M. (1988). The molecular genetics of Philadelphia chromosome-positive leukemias. *New Engl. J. Med.* **319**, 990–998.

Li, S., Janosch, P., Tanji, M., Rosenfeld, G. C., Waymire, J. C., Mischak, H., Kolch, W., and Sedivy, J. M. (1995). Regulation of Raf-1 kinase activity by the 14-3-3 family of proteins. *EMBO J.* **14**, 685–696.

Liu, D., Blenkowska, J., Petosa, C., Collier, R. J., Fu, H., and Liddington, R. (1995). Crystal structure of the zeta isoform of the 14-3-3 protein. *Nature (London)* **376**, 191–194.

Lu, G., de Vetten, N. C., Sehnke, P. C., Isobe, T., Ichimura, T., Fu, H., van Heusden, G. P., and Ferl, R. J. (1994). A single *Arabidopsis* GF14 isoform possesses biochemical characteristics of diverse 14-3-3 homologues. *Plant Mol. Biol.* **25**, 659–667.

Luo, Z. J., Zhang, X. F., Rapp, U., and Avuch, J. (1995). Identification of the 14-3-3 ζ domains important for self-association and Raf binding. *J. Biol. Chem.* **270**, 23681–23687.

Maehama, T., Takahashi, K., Ohoka, Y., Ui, M., and Katada, T. (1991). Identification of a botulinum C3-like enzyme in bovine brain that catalyzes ADP-ribosylation of GTP-binding proteins. *J. Biol. Chem.* **266**, 10062–10065.

Marais, R., Light, Y., Paterson, H. F., and Marshall, C. J. (1995). Ras recruits Raf-1 to the plasma membrane for activation by tyrosine phosphorylation. *EMBO J.* **14**, 3136–3145.

Maru, Y., and Witte, O. N. (1991). The BCR gene encodes a novel serine/threonine kinase activity within a single exon. *Cell* **67**, 459–468.

McWhirter, J. R., Galasso, D. L., and Wang, J. Y. J. (1993). A coiled-coil oligomerization domain of Bcr is essential for the transforming function of Bcr-Abl oncoproteins. *Mol. Cell. Biol.* **13**, 7587–7595.

Michaud, N. R., Fabian, J. R., Mathes, K. D., and Morrison, D. K. (1995). 14-3-3 is not essential for Raf-1 function: Identification of Raf-1 proteins that are biologically activated in a 14-3-3- and Ras-independent manner. *Mol. Cell. Biol.* **15**, 3390–3397.

Miyata, Y., and Yahara, I. (1992). The 90-kDa heat shock protein, HSP90, binds and protects casein kinase II from self-aggregation and enhances its kinase activity. *J. Biol. Chem.* **267,** 7042–7047.

Mochly-Rosen, D. (1995). Localization of protein kinases by anchoring proteins: A theme in signal transduction. *Science* **268,** 247–251.

Mochly-Rosen, D., Khaner, H., and Lopez, J. (1991a). Identification of intracellular receptor proteins for activated protein kinase C. *Proc. Natl. Acad. Sci. U.S.A.* **88,** 3997–4000.

Mochly-Rosen, D., Khaner, H., Lopez, J., and Smith, B. L. (1991b). Intracellular receptors for activated protein kinase C. *J. Biol. Chem.* **266,** 14866–14868.

Moore, B. W., and Perez, V. J. (1967). Specific acidic proteins of the nervous system. *In* "Physiological and Biochemical Aspects of Nervous Integration" (F. D. Carlson, ed.), pp. 343–349. Prentice-Hall, Englewood Cliffs, New Jersey.

Morgan, A., and Burgoyne, R. D. (1992a). Exo1 and Exo2 proteins stimulate calcium-dependent exocytosis in permeabilized adrenal chromaffin cells. *Nature (London)* **355,** 833–836.

Morgan, A., and Burgoyne, R. D. (1992b). Interaction between protein kinase C and Exo1 (14-3-3 protein) and its relevance to exocytosis in permeabilized adrenal chromaffin cells. *Biochem. J.* **286,** 807–811.

Morrison, D. (1994). 14-3-3: Modulators of signaling proteins? *Science* **266,** 56–57.

Moss, J., Tsai, S. C., and Vaughan, M. (1994). Activation of cholera toxin by ADP-ribosylation factors. *Methods Enzymol.* **235,** 640–647.

Muller, A. J., Young, J. C., Pendergast, A. M., Pondel, M., Landau, N. R., Littman, D. R., and Witte, O. N. (1991). BCR first exon sequences specifically activate the *BCR/ABL* tyrosine kinase oncogene of Philadelphia chromosome-positive human leukemias. *Mol. Cell. Biol.* **11,** 1785–1792.

Pallas, D. C., Fu, H., Haehnel, L. C., Weller, W., Collier, R. J., and Roberts, T. M. (1994). Association of polyomavirus middle tumor antigen with 14-3-3 proteins. *Science* **165,** 535–537.

Parker, P. J., Kour, G., Marais, R. M., Mitchell, F., Pears, C., Schaap, D., Stabel, S., and Webster, C. (1989). Protein kinase C—a family affair. *Mol. Cell. Endocrinol.* **65,** 1–11.

Pendergast, A. M., Muller, A. J., Havlik, M. H., Maru, Y., and Witte, O. N. (1991). BCR sequences essential for transformation by the *BCR-ABL* oncogene bind to the ABL SH2 regulatory domain in a non-phosphotyrosine-dependent manner. *Cell* **66,** 161–171.

Pendergast, A. M., Quilliam, L. A., Cripe, L. D., Bassing, C. H., Dai, Z., Li, N., Batzer, A., Rabun, K. M., Der, C. J., Schlessinger, J., and Gishizky, M. L. (1993). BCR-ABL-induced oncogenesis is mediated by direct interaction with the SH2 domain of the GRB-2 adaptor protein. *Cell* **75,** 175–185.

Prasad, G. L., Valverius, E. M., McDuffie, E., and Cooper, H. L. (1992). Complementary DNA cloning of a novel epithelial cell marker protein, HME1, that may be down-regulated in neoplastic mammary cells. *Cell Growth Differ.* **3,** 507–513.

Pratt, W. B., and Welsh, M. J. (1994). Chaperone functions of the heat shock proteins associated with steroid receptors. *Semin. Cell Biol.* **5,** 83–93.

Reuther, G. W., Fu, H., Cripe, L. D., Collier, R. J., and Pendergast, A. M. (1994). Association of the protein kinases cBcr and Bcr-Abl with proteins of the 14-3-3 family. *Science* **266,** 129–133.

Ridley, A. J., Self, A. J., Kasmi, F., Paterson, H. F., Hall, A., Marshall, C. J., and Ellis, C. (1993). Rho family GTPase activating proteins p 190, bcr, and rhoGAP show distinct specificites *in vitro* and *in vivo*. *EMBO J.* **12,** 5151–5160.

Robinson, K., Jones, D., Patel, Y., Martin, H., Madrazo, J., Martin, S., Howell, S., Elmore, M., Finnen, M. J., and Aitken, A. (1994). Mechanism of inhibition of protein kinase C by 14-3-3 isoforms. *Biochem. J.* **299,** 853–861.

Ron, D., Zannini, M., Levis, M., Wickner, R. B., Hunt, L. T., Graziani, G., Tronick, S. R., Aaronson, S. A., and Eva, A. (1991). A region of proto-*dbl* essential for its transforming activity shows sequence similarity to a yeast cell cycle gene, CDC 24, and the human breakpoint cluster gene, *bcr*. *New Biol.* **3,** 372–379.

Roth, D., Morgan, A., and Burgoyne, R. D. (1993). Identification of a key domain in annexin and 14-3-3 proteins that stimulate calcium-dependent exocytosis in permeabilized adrenal chromaffin cells. *FEBS Lett.* **320,** 207–210.

Roth, D., Morgan, A., Martin, H., Jones, D., Martens, G. J. M., Aitken, A., and Burgoyne, R. D. (1994). Characterization of 14-3-3 proteins in adrenal chromaffin cells and demonstration of isoform-specific phospholipid binding. *Biochem. J.* **301,** 305–310.

Sakariassen, K. S., Bolhuis, P. A., and Sixma, J. J. (1979). Human blood platelet adhesion to artery subendothelium. *Nature (London)* **279,** 636–638.

Sarafian, T., Aunis, D., and Bader, M. F. (1987). Loss of proteins from digitonin-permeabilized adrenal chromaffin cells essential for exocytosis. *J. Biol. Chem.* **262,** 16671–16676.

Shimizu, K., Kuroda, S., Yamamori, B., Matsuda, S., Kaibuchi, K., Yamauchi, T., Isobe, T., Irie, K., Matsumoto, K., and Takai, Y. (1994). Synergistic activation by Ras and 14-3-3 protein of a mitogen-activated protein kinase kinase kinase named Ras-dependent extracellular signal-regulated kinase kinase stimulator. *J. Biol. Chem.* **269,** 22917–22920.

Stokoe, D., Macdonald, S. G., Cadwallader, K., Symons, M., and Hancock, J. F. (1994). Activation of Raf as a result of recruitment to the plasma membrane. *Science* **264,** 1463–1467.

Sutherland, C., Alterio, J., Campbell, D. G., Le Bourdelles, B., Mallet, J., Haavik, J., and Cohen, P. (1993). Phosphorylation and activation of human tyrosine hydroxylase in vitro by mitogen-activated protein (MAP) kinase and MAP-kinase activated kinases 1 and 2. *Eur. J. Biochem.* **217,** 715–722.

Swanson, K. D., and Ganguly, R. (1992). Characterization of a *Drosophila melanogaster* gene similar to the mammalian genes encoding the tyrosine/tryptophan hydroxylase activator and protein kinase C inhibitor proteins. *Gene* **113,** 183–190.

Szyszka, R., Kramer, G., and Hardesty, B. (1989). The phosphorylation state of the reticulocyte 90-kDa heat shock protein affects its ability to increase phosphorylation of peptide initiation factor 2 α subunits by the heme-sensitive kinase. *Biochemistry* **28,** 1435–1438.

Tanji, M., Horwitz, R., Rosenfeld, G., and Waymire, J. C. (1994). Activation of protein kinase C by purified bovine brain 14-3-3: Comparison with tyrosine hydroxylase activation. *J. Neurochem.* **63,** 1908–1916.

Toker, A., Ellis, C. A., Sellers, L. A., and Aitken, A. (1990). Protein kinase C inhibitor proteins. *Eur. J. Biochem.* **191,** 421–429.

Toker, A., Sellers, L. A., Amess, B., Patel, Y., Harris, A., and Aitken, A. (1992). Multiple isoforms of a protein kinase C inhibitor (KCIP/14-3-3) from sheep brain. *Eur. J. Biochem.* **206,** 453–461.

van Heusden, G. P. H., Wenzel, T. J., Lagendijk, E. L., de Steensma, H. Y., and van den Berg, J. A. (1992). Characterization of the yeast *BMH1* gene encoding a putative protein homologous to mammalian protein kinase II activators and protein kinase C inhibitors. *FEBS Lett.* **302,** 145–150.

van Heusden, G. P. H., Griffiths, D. J. F., Ford, J. C., Chin-A-Woeng, T. F. C., Schrader, P.

A. T., Carr, A. M., and de Steensma, H. Y. (1995). The 14-3-3 proteins encoded by the *BMH1* and *BMH2* genes are essential in the yeast *Saccharomyces cerevisiae* and can be replaced by a plant homologue. *Eur. J. Biochem.* **229,** 45–53.

Voncken, J. W., van Schaick, H., Kaartinen, V., Deemer, K., Coates, T., Landing, B., Pattengale, P., Dorseuil, O., Bokoch, G. M., Groffen, J., and Heisterkamp, N. (1995). Increased neutrophil respiratory burst in *bcr*-null mutants. *Cell* **80,** 719–728.

Wang, W., and Shakes, D. C. (1994). Isolation and sequence analysis of a *Caenorhabditis elegans* cDNA which encodes a 14-3-3 homologue. *Gene* **147,** 215–218.

Wartmann, M., and Davis, R. J. (1994). The native structure of the activated Raf protein kinase is a membrane-bound multi-subunit complex. *J. Biol. Chem.* **269,** 6695–6701.

Wu, Y. N., Vu, N. D., and Wagner, P. D. (1992). Anti-(14-3-3 protein) antibody inhibits stimulation of noradrenaline (norepinephrine) secretion by chromaffin-cell cytosolic proteins. *Biochem. J.* **285,** 697–700.

Xiao, B., Smerdon, S. J., Jones, D. H., Dodson, G. G., Soneji, Y., Aitken, A., and Gamblin, S. J. (1995). Structure of a 14-3-3 protein and implications for coordination of multiple signaling pathways. *Nature (London)* **376,** 188–191.

Yamamori, B., Kuroda, S., Shimizu, K., Fukui, K., Ohtsuka, T., and Takai, Y. (1995). Purification of a Ras-dependent mitogen-activated protein kinase kinase kinase from bovine brain cytosol and its identification as a complex of B-Raf and 14-3-3 proteins. *J. Biol. Chem.* **270,** 11723–11726.

Yamauchi, T., Nakata, H., and Fujisawa, H. (1981). A new activator protein that activates tryptophan 5-monooxygenase and tyrosine 3-monooxygenase in the presence of Ca^{2+}-calmodulin-dependent protein kinase. *J. Biol. Chem.* **256,** 5404–5409.

Zupan, L. A., Steffens, D. L., Berry, C. A., Landt, M., and Gross, R. W. (1992). Cloning and expression of a human 14-3-3 protein mediating phospholipolysis. *J. Biol. Chem.* **267,** 8707–8710.

Physiological Roles for Parathyroid Hormone-Related Protein: Lessons from Gene Knockout Mice

ANDREW C. KARAPLIS* AND HENRY M. KRONENBERG[†]

Division of Endocrinology, McGill University, Sir Mortimer B. Davis–Jewish General Hospital, and Lady Davis Research Institute, Montréal, Canada H3T 1E2 and †Endocrine Unit, Massachusetts General Hospital, Harvard Medical School, Boston, Massachusetts 02114

I. Introduction
II. Strategy for Generating PTHrP-Negative Mice
   A. Disruption of PTHrP Gene in Embryonic Stem Cells
   B. Generation of Chimeric Mice and Germline Transmission
III. The PTHrP-Negative Phenotype
   A. Perinatal Lethality with Disruption of PTHrP Gene
   B. Skeletal Abnormalities in the PTHrP-Negative Phenotype
   C. Histological Examination of the Mutant Skeleton
   D. PTHrP and PTH–PTHrP Receptor Expression in Fetal Tibiae
IV. Summary
References

I. INTRODUCTION

Parathyroid hormone-related peptide (PTHrP) was originally identified because of its production by tumors associated with humoral hypercalcemia of malignancy (for review, see Broadus and Stewart, 1994). This common paraneoplastic syndrome is characterized by hypercalcemia and hypophosphatemia, biochemical abnormalities often associated with oversecretion of parathyroid hormone (PTH), the major peptide regulator of calcium homeostasis. The homology of the amino-terminal region of PTHrP with the corresponding domain of PTH, and the resultant capacity of both molecules to interact with a common receptor (Jüppner et al., 1991; Abou-Samra et al., 1992), appear to account for the ability of PTHrP to mimic many of the effects of PTH on calcium and phosphate homeostasis and on skeletal turnover. However, unlike PTH, whose synthesis is restricted to the parathyroid glands, PTHrP is produced in a number of normal adult and fetal tissues (Broadus and Stewart, 1994), suggesting that it may play a broader role than simply as a hypercalcemia-inducing oncoprotein.

The human PTHrP gene is a complex transcriptional unit composed of six exons (Mangin *et al.*, 1989, 1990b). This gene uses at least three promoters (Vasavada *et al.*, 1993) and, via alternate splicing, it gives rise to three major isoforms of the mature peptide, of 139, 141, and 173 amino acids. The single transcript for rat (Thiede and Rodan, 1988; Yasuda *et al.*, 1989) and mouse (Mangin *et al.*, 1990a) PTHrP encodes the 141-residue isoform (139 amino acids in the mouse because of a 2–amino acid deletion), while two chicken transcripts encode proteins of 139 and 141 amino acids. Human, rat, mouse, and chicken PTHrP share marked amino acid sequence identity throughout the amino-terminal and midregion portions of the protein but diverge beyond residue 112. The striking conservation of the protein sequence throughout this large evolutionary range suggests that PTHrP must have an essential role in normal physiology.

In contradistinction to the well-characterized functions of PTH, a physiological role for PTHrP is more difficult to define. Given that its synthesis and secretion are not limited to a defined tissue, the classical endocrine approach of tissue ablation, used in defining a physiological role for PTH, is not possible. Therefore, studies directed toward defining a function for PTHrP have utilized primarily *in vitro* systems. Numerous such studies suggest that PTHrP acts as a paracrine, autocrine, and at times endocrine factor. Several important roles have been proposed for PTHrP, including smooth muscle relaxation, neurotransmission, transplacental calcium transport, and control of tissue growth and differentiation (see review by Broadus and Stewart, 1994, and references therein). Nevertheless, little is known of the definitive role of PTHrP in the development and function of the organism *in vivo*. Although these elegant studies have provided an array of hypotheses regarding the physiological role of this protein, their conclusions must be confirmed *in vivo* to establish validity.

The advent of transgenic mouse technology has made it possible to address some of these issues in the intact organism. Using the traditional transgenic approach, in which a new gene is added to the mouse genome and associated abnormalities are examined, PTHrP overexpression in skin (Wysolmerski *et al.*, 1994) and mammary tissue (Wysolmerski *et al.*, 1993) of mice has suggested a role for this protein in cellular proliferation and differentiation. These studies, however, can only suggest the physiological actions of normal, rather than high, levels of PTHrP in skin and breast.

Advances in transgenic technology have made it possible now for genes to be selectively removed from the mouse genome (Capecchi, 1989). With this approach, known as gene targeting by homologous recombination, mouse strains carrying a specific recessive mutation of interest can be generated (Bradley, 1987) and studied for associated abnormalities. This gene ablation technology clearly complements the classical endocrine approach of tissue ablation and is ideally suited for studying the role of paracrine–autocrine factors such as PTHrP *in vivo*. The exciting potential for such genetic manipulation served as the impetus for employing this approach to generate mice missing PTHrP, thereby revealing *in vivo* functions of this protein.

II. Strategy for Generating PTHrP-Negative Mice

Our gene ablation strategy focused on the removal of the PTHrP gene's major coding exon (exon IV) from the murine genome. Such a deletion would be expected to make synthesis of PTHrP impossible (Karaplis *et al.*, 1994).

A. Disruption of PTHrP Gene in Embryonic Stem Cells

The murine PTHrP gene was isolated from a phage BALB/c mouse genomic library following plaque hybridization with a rat PTHrP cDNA probe. Three clones were identified and plaque purified, and one clone was used for construction of the targeting vector. This clone contained sequences for exons IV and V of the murine PTHrP gene, encoding the mature 139-amino acid protein and the 3' untranslated region.

To target the PTHrP gene, a sequence replacement vector was designed using the pPNT plasmid (Tybulewicz *et al.*, 1991) and fragments of the cloned PTHrP gene derived from phage DNA (Fig. 1). Approximately 7 kb of PTHrP genomic DNA were used for the construction of this vector. A 3.5-kb genomic fragment from sequences 5' to exon IV and a 3.6-kb fragment 3' to exon IV were inserted into pPNT to generate the targeting vector. A double crossing-over event would substitute the bacterial neomycin resistance gene (neo^r)-selectable marker for endogenous sequences containing exon IV. Since this exon comprises nearly all of the coding region of the mature protein (amino acids 1–137), the disruption would be predicted to result in a recombinant allele incapable of directing the synthesis of a functional PTHrP protein.

The targeting vector was linearized and introduced into recipient D3 embryonic stem (ES) cells (Doetschman *et al.*, 1985) by electroporation. Enrichment for mutant ES cells with one damaged PTHrP allele was achieved using positive–negative selection (Capecchi, 1989) with

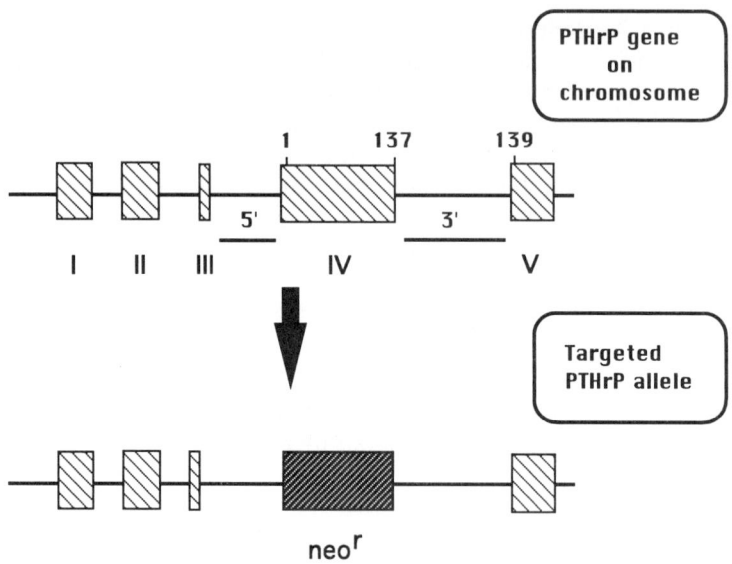

FIG. 1. Schematic representation of the murine PTHrP gene and the resultant targeted allele following homologous recombination. 5′ and 3′ represent genomic DNA fragments used for the construction of the targeting vector; neo^r, bacterial neomycin resistance gene.

the drugs G418 and 1-(2-deoxy, 2-fluoro-β-D-arabinofuranosyl)-5-iodouracil (FIAU). The enrichment attributed to FIAU selection was approximately 17-fold.

After selection, colonies resistant to both drugs were clonally expanded, and identification of clones with a disrupted PTHrP gene was accomplished by Southern blot analysis. Ninety-five doubly resistant clones were analyzed and six independent homologous integration events were identified. Thus, homologous recombination was observed in ~6% of all doubly resistant clones. The absence of additional random copies of the replacement vector was verified by Southern blot analysis using neo^r gene sequences as a probe.

B. Generation of Chimeric Mice and Germline Transmission

Cells from three independent mutant clones were introduced by microinjection into blastocysts of C57BL/6 mice to generate chimeras. Two of the clones resulted in extensively chimeric male mice in the progeny who were able to transmit the mutated PTHrP allele through the germline when mated to C57BL/6 females. Mice heterozygous for the targeted allele were identified by Southern blot analysis of tail genomic DNA. As expected, 50% of the agouti offspring carried the mutated PTHrP allele. (Agouti is a dominant hair color trait carried by the ES cell line but not by the host blastocyst line.) These animals had no readily distinct phenotype when compared to their wild-type litter mates, grew normally, and were able to breed successfully.

III. The PTHrP-Negative Phenotype

A. Perinatal Lethality with Disruption of PTHrP Gene

To define the consequences associated with complete absence of PTHrP, heterozygous mice were interbred and offspring were genotyped 3 weeks after birth. However, no homozygous PTHrP mutants were identified, suggesting that absence of PTHrP was lethal. When mice were examined a day before anticipated parturition, in contrast, 20% of them were homozygous for the PTHrP gene deletion. Thus, loss of the PTHrP gene is not required for viability throughout fetal life. From witnessing spontaneous, full-term births, it became evident that mice missing PTHrP died immediately after parturition. These animals were hypotonic and made feeble attempts to breathe. Thus, PTHrP-negative homozygous mice survived until birth but died immediately after delivery, possibly from respiratory failure. Occasional animals live for a few hours before dying.

These animals exhibited size and body weight similar to those of their wild-type and heterozygous litter mates. Nevertheless, they had a distinct phenotype characterized by domed skull, short snout and mandible, protruding tongue, and disproportionately short limbs (Fig. 2), indicative of a form of osteochondrodysplasia.

B. Skeletal Abnormalities in the PTHrP-Negative Phenotype

To further assess these phenotypic abnormalities, bones of 18.5-day mutant fetuses were examined following staining with alizarin red S, which stains the calcified skeleton. A multitude of abnormalities became apparent in these preparations, with the most striking feature being the advanced mineralization evident throughout the endochondral (arising from cartilaginous mold) skeleton (axial as well as appendicular). In sharp contrast, no abnormalities were noted in skeletal structures that develop entirely by intramembranous ossification (without replacing a cartilaginous mold).

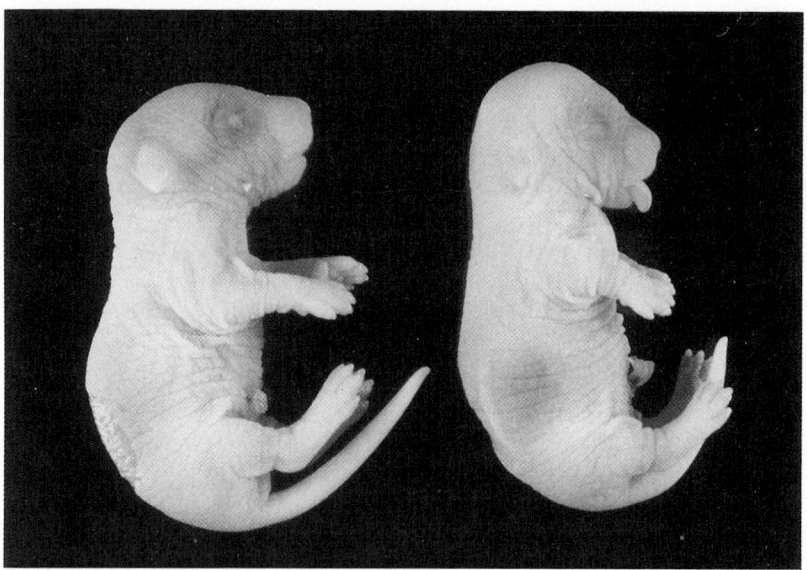

FIG. 2. Wild-type (left) and PTHrP-negative (right) fetuses at 18.5 days of gestation. The chondrodysplastic phenotype is characterized by the domed cranium, short mandible, protruding tongue, and greatly foreshortened extremities. (From Karaplis et al., 1994.)

In the homozygous mutants there was a striking paucity of the normally unstained cartilaginous portions of the ribs and sternum (Fig. 3). As a consequence, the ribs were short and splayed, and the sternum was also shorter than normal; these changes resulted in a narrow, bell-shaped thoracic cage with markedly diminished cross-sectional area. This deformity, by causing respiratory compromise, may have contributed to the perinatal lethality of the mutants. Similar abnormalities are often the cause of mortality associated with lethal forms of human osteochondrodysplasias.

The appendicular endochondral skeleton also revealed developmental abnormalities arising in the absence of PTHrP, in that the long bones were shorter and thicker than those from wild-type litter mates, and had marked deformities (see later). Again, there was a striking diminution in the cartilaginous nonstaining portions of the bones, as indicated by the diminished space between calcified–ossified segments. The absence of PTHrP is therefore associated with premature maturation and ossification of the mutant cartilaginous skeleton and, hence, inappropriate acceleration of the normal endochondral ossification process.

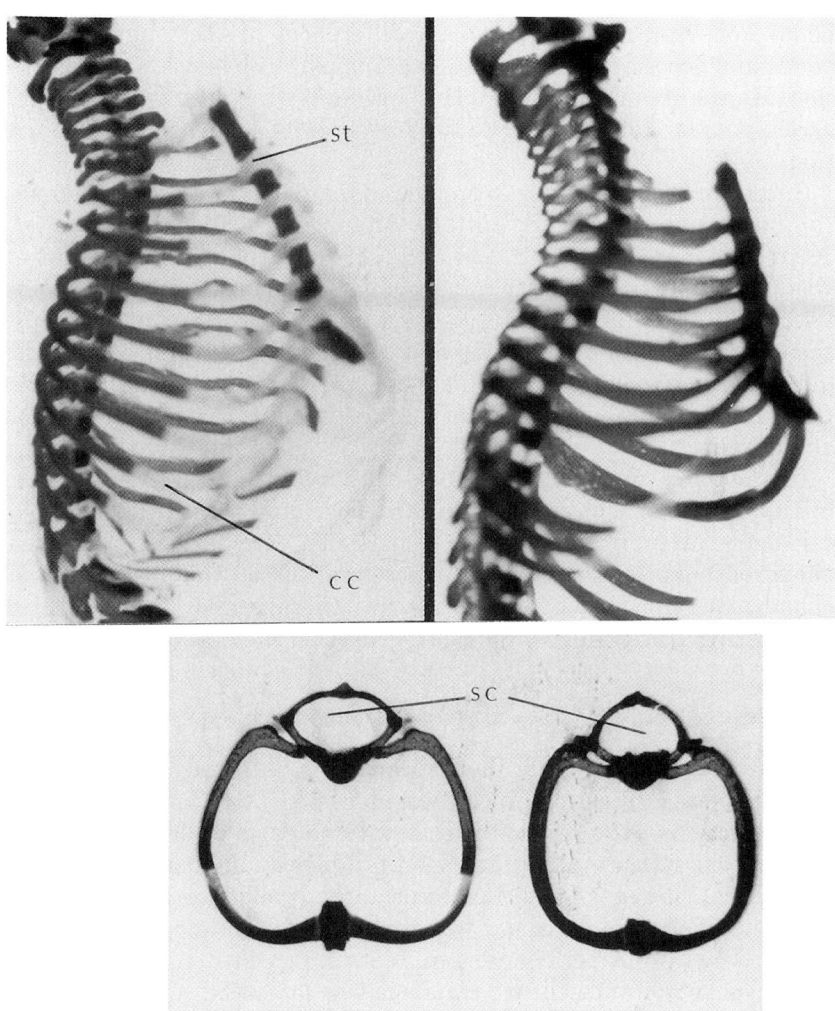

FIG. 3. (Top) Thoracic cage of wild-type (left) and mutant (right) fetuses stained with alizarin red S. The normal nonstaining cartilaginous portions of the ribs and sternum were homogeneously stained with the dye in the mutant specimen, suggesting diminished cartilaginous growth and premature mineralization. (Bottom) Cross section of the wild-type (left) and mutant (right) rib cage. St, sternebrae; cc, costal cartilage; sc, spinal canal. (From Karaplis et al., 1994.)

Studies were conducted to define the ontogeny of the skeletal abnormalities associated with PTHrP deficiency. The process of mesenchymal condensation to form the initial cartilaginous blastemata of long bones appeared normal. However, differences in the length of long bones and associated deformities were apparent by day 14.5 of gestation. Hence, the absence of PTHrP influences the development of the cartilaginous skeleton prior to the initiation of endochondral ossification.

Since the PTHrP gene is expressed in many fetal tissues in a temporally and spatially specific pattern (Moniz et al., 1990; Campos et al., 1991; Senior et al., 1991; Stolpe et al., 1993), a thorough examination was undertaken to identify abnormalities in nonskeletal organs and tissues in the PTHrP homozygous mutants. Gross examination of these animals was unremarkable except for the skeletal deformities already described. Moreover, histological assessment of multiple sections from a number of mutant tissues failed to identify any other abnormalities.

In summary, mice homozygous for the mutated PTHrP gene die at birth, perhaps of asphyxia. These animals exhibit a multitude of skeletal deformities phenotypically consistent with a form of osteochondrodysplasia. Inappropriate acceleration of the normal endochondral ossification process underlies the observed abnormalities, implicating a critical role for PTHrP in fetal skeletal development.

C. Histological Examination of the Mutant Skeleton

To define more precisely the cellular basis of the advanced skeletal calcification–ossification process noted in the PTHrP homozygous mutants, sections were prepared from various parts of the skeletons of 18.5-day-old fetuses. Shown in Fig. 4 is a longitudinal section through the ventral rib cage of a normal fetus at the level of the fifth rib. Here, the zone of the rib cartilage was composed of hyaline cartilage surrounded by fibrous perichondrium. In striking contrast, in the PTHrP homozygous mutants, the rib cartilage was composed of more differentiated (hypertrophic) chondrocytes encased by perichondrial bone, as indicated by the presence of characteristic bony matrix, active osteoblasts, and osteocytes. These findings, also observed in other areas of abnormal alizarin red S staining, were indicative of widespread premature maturation and ossification of the mutant cartilaginous skeleton as a consequence of the absence of PTHrP.

Because abnormal growth of long bones was an outstanding feature of the PTHrP homozygous mutant phenotype, studies also focused on the morphology of epiphyseal cartilage and growth plates (Amizuka et al., 1994). Histological examination of mutant growth plates of long bones revealed marked quantitative and qualitative abnormalities

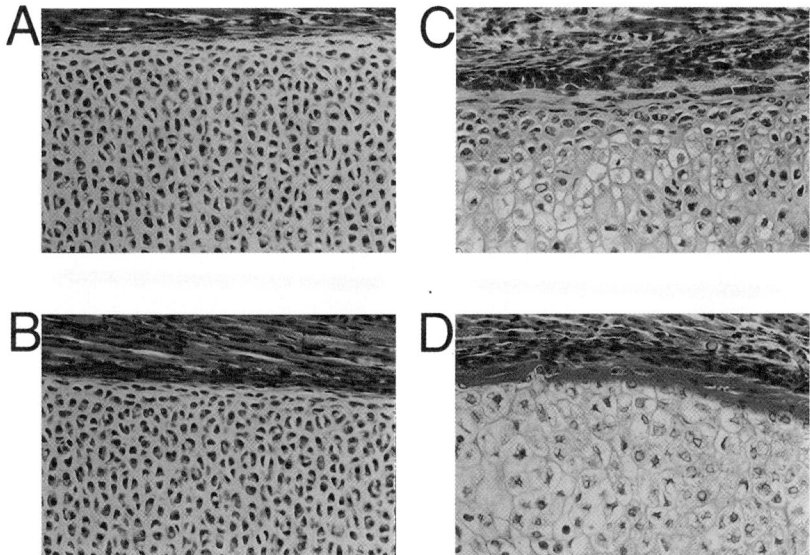

FIG. 4. Histological examination of wild-type (A, B) and mutant (C, D) rib cartilage. In the absence of PTHrP, hyaline cartilage is altered to hypertrophic chondrocytes with associated perichondrial ossification, indicative of inappropriate differentiation program. A and C stained with hematoxylin and eosin; B and D stained with van Gieson's trichrome. (From Karaplis et al., 1994.)

(Fig. 5). Primarily, there was a marked reduction in the height of the cartilaginous zones, particularly the zone of resting and proliferating chondrocytes. In the hypertrophic zone of wild-type mice, all chondrocytes became enlarged. However, in the mutant hypertrophic zone, clusters of nonhypertrophic chondrocytes persisted among the cells in the hypertrophic zone. Hence, in the mutant specimen, the transition zone from proliferative to hypertrophic chondrocytes began at various levels with no clear junction in between.

By morphological and biochemical criteria, these nonhypertrophic cells resembled resting or proliferative chondrocytes. Ultrastructurally, they contained distended cisternae of rough endoplasmic reticulum, a moderately well-developed Golgi apparatus, and secretory vesicles. An unusual but characteristic finding in the PTHrP-null mice was the presence of large dense masses of glycogen in the cytoplasm of the nonhypertrophic chondrocytes. Following in vivo labeling with [^{3}H] thymidine, these cells showed numerous silver grains over the nuclei indicative of active DNA synthesis and, thus, proliferation. In contrast, nonhypertrophic chondrocytes exhibited intense alkaline phos-

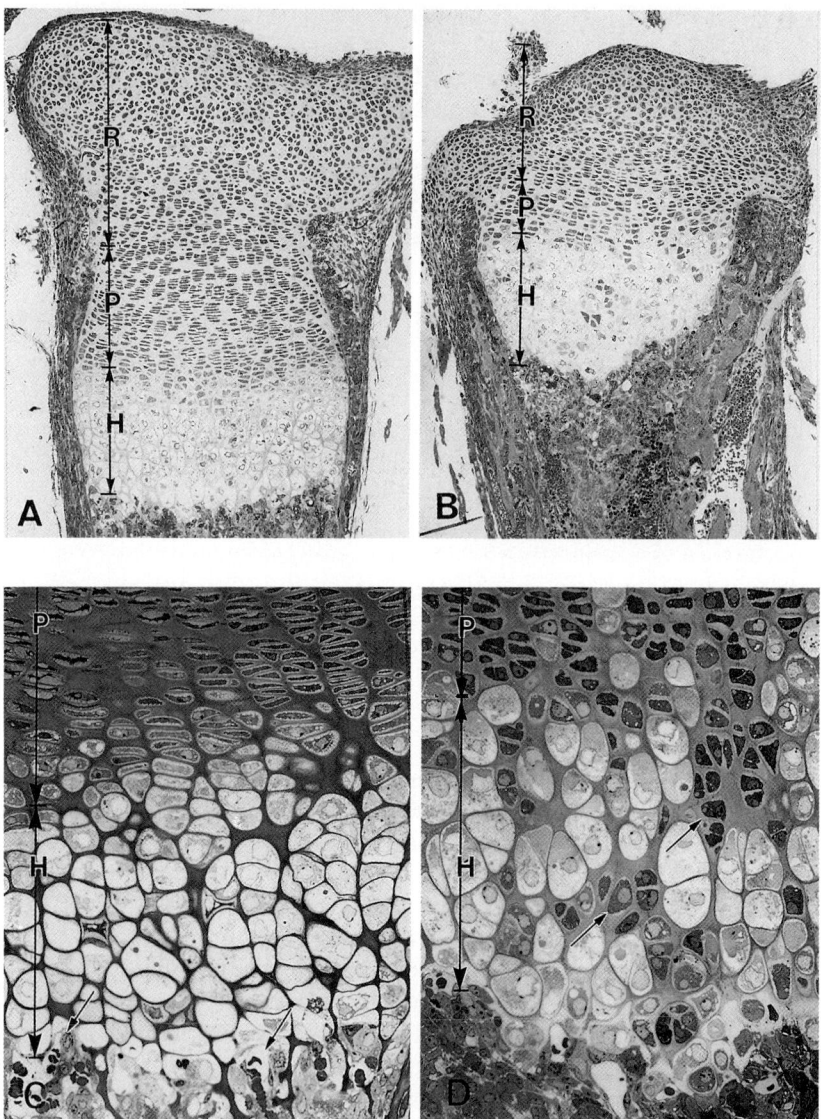

FIG. 5. Histological differences in the growth plates of tibiae of wild-type (A,C) and mutant (B,D) mice. A and B show the resting (R) and proliferating (P) zones being much shorter in the mutant, while the zones of hypertrophy were of similar size. C and D are photomicrographs of the zone of hypertrophy. In the mutant, nonhypertrophic chondrocytes (arrows) within the zone disrupt the formation of longitudinal columns. (From Amizuka et al., 1994. Reproduced from J. Cell Biol. **126,** 1611–1623 by copyright permission of The Rockefeller University Press.)

phatase activity on the cell membrane, similar to hypertrophic chondrocytes.

The persistence of clusters of nonhypertrophic chondrocytes up to the metaphyseal side of the epiphyseal plate distorted the longitudinal columns of hypertrophic chondrocytes, resulting in the absence of longitudinal septa of cartilage matrix. In the zone of calcifying cartilage in wild-type mice, calcification, in the form of electron-dense hydroxyapatite crystals, was uniformly distributed in the longitudinal intercolumnar septae of cartilage matrix between hypertrophic chondrocytes (Fig. 6). In the homozygous mutants, calcification occurred unevenly, with hydroxyapatite crystals deposited only in the vicinity of hypertrophic chondrocytes. Calcification was not evident adjacent to nonhypertrophic cells located in this zone, accounting for distorted chondrocyte columns and sporadic distribution of calcified cartilage. Consequently, in the metaphysis, bone was deposited on the irregular and sparse scaffold of calcified cartilage. The resulting mixed spicules did not parallel the longitudinal axis of the bone and were therefore inappropriate for bone elongation, resulting in foreshortened and deformed long bones characteristic of the PTHrP-negative chondrodysplastic phenotype.

D. PTHrP and PTH–PTHrP Receptor Expression in Fetal Tibiae

In view of the profound alterations observed in the mutant skeleton, the expression of PTHrP and its amino-terminal receptor was examined in the tibiae of 18.5-day-old wild-type fetuses (Fig. 7). PTHrP-immunopositive chondrocytes were observed primarily in the thin zone of transition between proliferative and hypertrophic zones. Immunoreactive chondrocytes were also seen in the resting and proliferative zones. However, PTHrP immunoreactivity was not observed in the hypertrophic zone nor in the articular or subarticular regions at the head of the tibia. In contrast to the localization of PTHrP immunoreactivity, the major site of PTHrP mRNA was the perichondrium (Lee et al., 1994).

Messenger RNA encoding the PTH–PTHrP receptor was distributed mainly in the proliferative zone of the epiphyseal growth plate, including the upper region of the hypertrophic zone, but not in the lower hypertrophic zone. This pattern of expression was unchanged in the homozygous mutant animals.

In summary, in the mutant epiphysis, the proliferation and orderly transition from proliferative to hypertrophic chondrocytes is clearly altered. The synthesis of PTHrP within the developing bone suggests

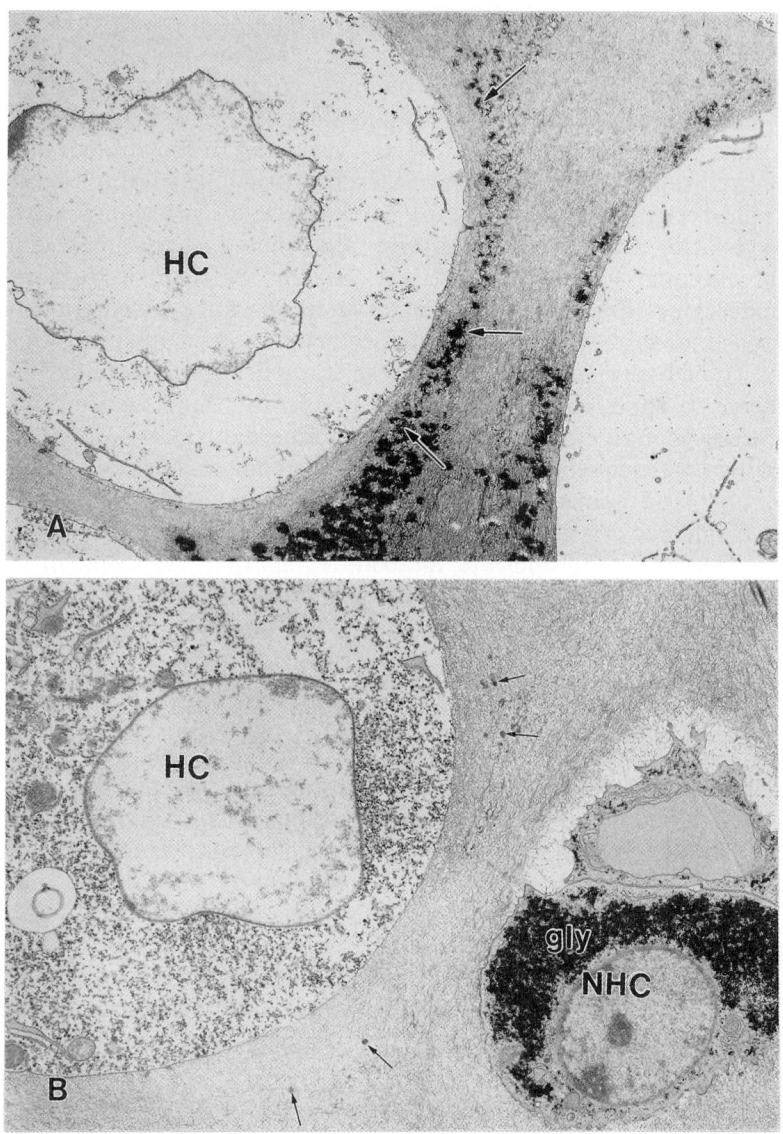

FIG. 6. Electron micrograph of the zone of calcifying cartilage. In the wild-type mouse (A), deposition of hydroxyapatite crystals is seen as dense particles (arrows) adjacent to hypertrophic chondrocytes (HC). In the mutant mouse (B), there is deficient calcification adjacent to nonhypertrophic chondrocytes (NHC). Note glycogen (gly) accumulation in nonhypertrophic cells. (From Amizuka et al., 1994. Reproduced from J. Cell Biol. **126,** 1611–1623 by copyright permission of the Rockefeller University Press.)

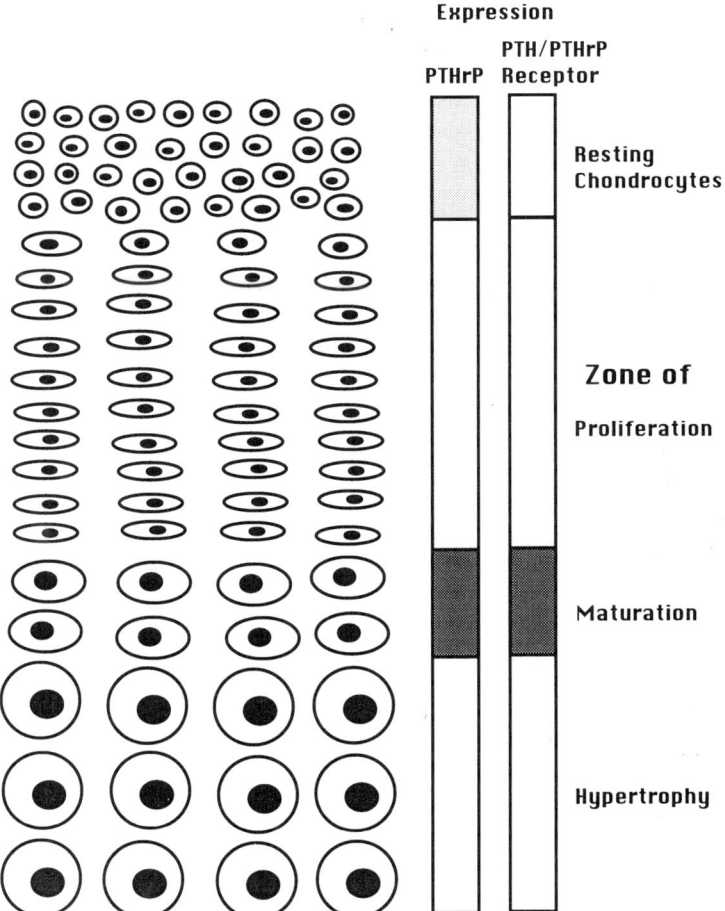

FIG. 7. Schematic representation of PTHrP immunoreactivity and PTH–PTHrP receptor mRNA expression in the normal fetal epiphysis. Degree of shading correlates with level of expression.

that locally produced PTHrP, perhaps acting on the PTH–PTHrP receptor, regulates these processes. The striking decrease in the extent of the zone of proliferating chondrocytes, with the differentiation of chondrocytes into hypertrophic chondrocytes much closer than normal to the articular base of the growth plate, suggests abnormalities in chondrocyte proliferation, differentiation, or both. The loss of proliferating chondrocytes might reflect a role for PTHrP in stimulating proliferation of these cells. Alternatively, the major action of PTHrP

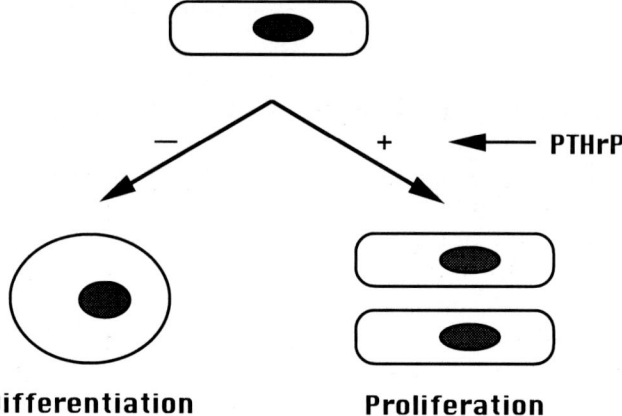

FIG. 8. The influence of PTHrP on chondrocyte biology. The PTHrP-negative phenotype is interpreted as shown in this figure. PTHrP influences the rate of proliferation and degree of differentiation of growth plate chondrocytes. In the absence of PTHrP, proliferation decreases and there is inappropriate and premature differentiation to hypertrophic chondrocytes.

could be to delay the differentiation of chondrocytes into nondividing, hypertrophic chondrocytes. This delay would allow more extensive proliferation of less differentiated chondrocytes. Of course, these alternative possible actions of PTHrP are not mutually exclusive (Fig. 8).

IV. Summary

A null mutation in the PTHrP gene produces profound abnormalities in endochondral bone formation *in vivo*. The role of PTHrP as a modulator of the chondrocytic proliferation and differentiation program is evident in the alterations that occur in its absence in the temporal and spatial sequence of chondrocyte development and subsequent endochondral bone formation that is necessary for normal bone elongation. These actions of PTHrP are probably responsible for the delay in chondrocyte development seen in Jansen osteochondrodystrophy, a disease caused by ligand-independent activation of the PTH–PTHrP receptor (Schipani *et al.*, 1995). Furthermore, these conclusions have been corroborated by the observation that chondrocyte-specific overexpression of PTHrP causes a profound delay in the developmental program of chondrocyte differentiation and endochondral ossification (Weir et al., 1995).

The morphological abnormalities in the knockout mice were limited to the skeletal system, despite the widespread production of PTHrP during fetal development. At this point, one can only speculate about the limited tissue distribution of the abnormalities. It is possible, for example, that other gene products, such as PTH, can compensate for the loss of PTHrP in some tissues. Alternatively, possible abnormalities in proliferation and differentiation may be present but morphologically subtle. A molecular assessment of these possible actions may well reveal more widespread effects of PTHrP.

ACKNOWLEDGMENTS

We are gratefully indebted to a number of colleagues and collaborators who have contributed to this work. In particular we thank Drs. D. Goltzman, N. Amizuka, J. E. Henderson, and H. Warshawsky at McGill, for sharing artwork and ideas; Drs. B. Lanske and C. S. Kovaks at the Endocrine Unit, Massachusetts General Hospital, for sharing unpublished information; Dr. A. Luz at the Institut für Pathologie, GSF München; and Dr. R. C. Mulligan at the Whitehead Institute, Massachusetts Institute of Technology. This work was supported in part by the Medical Research Council (MRC) of Canada (A. C. Karaplis) and National Institutes of Health grant DK47038 (H. M. Kronenberg); A. C. Karaplis is a recipient of an MRC Scholarship.

REFERENCES

Abou-Samra A.-B., Jüppner, H., Force, T., Freeman, M. W., Kong, X. F., Schipani, E., Ureña, P., Richards, J., Bonventre, J. V., Potts, J. T. Jr., Kronenberg, H. M., and Segre, G. V. (1992). Expression cloning of a common receptor for parathyroid hormone and parathyroid hormone-related peptide from rat osteoblast-like cells: A single receptor stimulates intracellular accumulation of both cAMP and inositol trisphosphates and increases intracellular free calcium. *Proc. Natl. Acad. Sci. U.S.A.* **89**, 2732–2736.

Amizuka, N., Warshawsky, H., Henderson, J. E., Goltzman, D., and Karaplis, A. C. (1994). Parathyroid hormone-related peptide-depleted mice show abnormal epiphyseal cartilage development and altered endochondral bone formation. *J. Cell Biol.* **126**, 1611–1623.

Bradley, A. (1987). Production and analysis of chimeric mice. In "Teratocarcinomas and Embryonic Stem Cells: A Practical Approach" (E. J. Robertson, ed.), p. 113. IRL Press, Washington, DC.

Broadus, A. E., and Stewart, A. F. (1994). Parathyroid hormone-related protein. Structure, processing, and physiological actions. In "The Parathyroids" (J. P. Bilezikian, R., Marcus, and M. A. Levine, eds.), p. 259. Raven Press, New York.

Campos, R. V., Asa, S. L., and Drucker, D. J. (1991). Immunocytochemical localization of parathyroid hormone-like peptide in the rat fetus. *Cancer Res.* **51**, 6351–6357.

Capecchi, M. R. (1989). Altering the genome by homologous recombination. *Science* **244**, 1288–1292.

Doetschman, T. C., Eistetter, H., Katz, M., Schmidt, W., and Kemler, R. (1985). The *in vitro* development of blastocyst-derived embryonic stem cell lines: Formation of visceral yolk sac, blood islands and myocardium. *J. Embryol. Exp. Morphol.* **87**, 27–45.

Jüppner, H., Abou-Samra, A.-B., Freeman, M., Kong, X.-F., Schipani, E., Richards, J., Kolakowski, L. F. Jr., Hock, J., Potts, J. T. Jr., Kronenberg, H. M., and Segre, G. V. (1991). A G protein-linked receptor for parathyroid hormone and parathyroid hormone-related peptide. *Science* **254**, 1024–1026.

Karaplis, A. C., Luz, A., Glowacki, J., Bronson, R. T., Tybulewicz. V. L. J., Kronenberg, H. M., and Mulligan, R. C. (1994). Lethal skeletal dysplasia from targeted disruption of the parathyroid hormone-related peptide gene. *Genes Dev.* **8**, 277–289.

Lee, K., Karaplis, A., Deeds, J., Lanske, B., Kronenberg, H. M., and Segre, G. V. (1994). Molecular analysis of abnormal endochondral bone formation in PTHrP-less mice. *J. Bone Miner. Res.* **9** (Suppl 1), S159.

Mangin, M., Ikeda, K., Dreyer, B. E., and Broadus, A. E. (1989). Isolation and characterization of the human parathyroid hormone-like peptide gene. *Proc. Natl. Acad. Sci. U.S.A.* **86**, 2408–2412.

Mangin, M., Ikeda, K., and Broadus, A. E. (1990a). Structure of the mouse gene encoding parathyroid hormone-related peptide. *Gene* **95**, 195–202.

Mangin, M., Ikeda, K., Dreyer, B. E., and Broadus, A. E. (1990b). Identification of an upstream promoter of the human parathyroid hormone-related peptide gene. *Mol. Endocrinol.* **4**, 851–858.

Moniz, C., Burton, P. B. J., Malik, A. N., Dixit, M., Banga, J. P., Nicolaides, K., Quirke, P., Knight, D. E., and McGregor, A. M. (1990). Parathyroid hormone-related peptide in normal human fetal development. *J. Mol. Endocrinol.* **5**, 259–266.

Schipani, E., Kruse, K., and Jüppner, H. (1995). A constitutively active PTH/PTHrP receptor in Jansen type metaphyseal chondrodysplasia. *Science* **268**, 98–100.

Senior, P. V., Heath, D. A., and Beck, F. (1991). Expression of parathyroid hormone-related protein mRNA in the rat before birth: Demonstration by hybridization histochemistry. *J. Mol. Endocrinol.* **6**, 281–290.

Stolpe, A. v. d., Karperien, M., Lowik, C. W. G. M., Jüppner, H., Segre, G. V., Abou-Samra, A.-B., Laat, S. W. d., and Defize, L. H. K. (1993). Parathyroid hormone-related peptide as an endogenous inducer of parietal endoderm differentiation. *J. Cell Biol.* **120**, 235–243.

Thiede, M. A., and Rodan, G. A. (1988). Expression of a calcium-mobilizing parathyroid hormone-like peptide in lactating mammary tissue. *Science* **242**, 278–280.

Tybulewicz, V. L. J., Crawford, C. E., Jackson, P. K., Bronson, R. T., and Mulligan, R. C. (1991). Neonatal lethality and lymphopenia in mice with a homozygous disruption of the c-abl proto-oncogene. *Cell* **65**, 1153–1163.

Vasavada, R. C., Wysolmerski, J. J., Broadus, A. E., and Philbrick, W. M. (1993). Identification and characterization of a GC-rich promoter of the human parathyroid hormone-related peptide gene. *Mol. Endocrinol.* **7**, 273–282.

Weir, E., Philbrick, W., Neff, L., Amling, M., Baron, R., and Broadus, A. (1995). Targeted overexpression of parathyroid hormone-related peptide in chondrocytes causes skeletal dysplasia and delayed osteogenesis. *J. Bone Miner. Res.* **10**, (Suppl 1), S157.

Wysolmerski, J., Daifotis, A., Broadus, A., Milstone, L., and Philbrick, W. (1993). Overexpression of PTHrP in transgenic mice results in breast hypoplasia. *J. Bone Miner. Res.* **8**, (Suppl. 1), S149.

Wysolmerski, J. J., Broadus, A. E., Zhou, J., Fuchs, E., Milstone, L., and Philbrick, W. M. (1994). Overexpression of parathyroid hormone-related protein in skin of transgenic mice interferes with hair follicle development. *Proc. Natl. Acad. Sci. U.S.A.* **91,** 1133–1137.

Yasuda, T., Banville, D., Rabbani, S. A., Hendy, G. N., and Goltzman, D. (1989). Rat parathyroid hormone-like peptide: Comparison with the human homologue and expression in malignant and normal tissue. *Mol. Endocrinol.* **3,** 518–525.

Index

A

ACTH, see Adrenocorticotropic hormone
Ad4BP, see Steroidogenic factor-1
Adenylate cyclase
  coupling of peptide hormone receptors to, 130
  estrogen activation of, 101
ADP, 14-3-3 protein and ribosylation function, 155–156
Adrenocorticotropic hormone
  elevation of steroidogenic enzyme levels in bovine adrenocortical cells by, 136
  lack of cortisol in bovine fetal blood in absence of, 143
Aging, see Elderly
Agonists, versus ovarian steroid receptor antagonists, 110–117
Annexins
  effects on osteoclast-like cell formation and bone resorption, 84
  14-3-3 protein sequence homology to, 150
Antagonists, see also Antihormones
  ovarian steroid receptor, 109–111, 112–117
Anterior pituitary hormones, see Peptide hormones
Antiestrogens, 117–120
Antihormones, ovarian steroid receptor
  estrogen receptor function antagonism, 117–120
  overview, 109–110
  progesterone receptor function antagonism, 112–117
Antioxidants, see also specific antioxidants
  immune response and
    in cigarette smoking, 53–55
    deficiency, 41–42
    effects in elderly, 46–53
    rheumatoid arthritis, 43–44
    supplementation in elderly, 45–46
  increased need under oxidative stress conditions, 55–56
  LDL oxidation and
    controversy over in vitro versus in vivo effects, 20–21, 26–27
    initial suppression by endogenous, 9
    sequence of consumption in LDL peroxidation, 15
    supplementation trials in vivo, 21–27
    types associated with LDL, 5, 6t
  micronutrients acting as, 36–37
Antiprogestins, 112–117
Apolipoprotein B, in oxidation of human LDL, 5–6, 9
Ascorbic acid, see Vitamin C
Atherosclerosis
  effects of antioxidant supplementation, 21–27
  pathogenesis
    mechanism, 1–2
    oxidative modification hypothesis, 26
    promotion versus inhibition by vitamin E supplementation, 20–21
    role of oxidized LDL in, 2–5
    vitamin C inhibition of lesion development, 9–14
Atherosclerotic plaques, 2
Autocrine–paracrine factors, stimulation of osteoclasts by, 80–84

B

Bcr-Abl oncogene, 14-3-3 protein regulation of, 165–166
Bone
  abnormalities in PTHrP-negative mice
    histological examination results, 184–187
    skeletal structure, 181–184
  stages of in vitro development, 64–65
Bone cell differentiation, see also PTHrP-negative mice, phenotype

Bone cell differentiation (*Continued*)
  osteoblasts
    general characteristics, 64
    proliferation and differentiation factors, 65–76
    systems for *in vitro* study, 64–65
  osteoclasts
    autocrine–paracrine factors stimulating, 80–84
    general characteristics, 76–77
    hormones affecting function and formation of, 78–80
    local inhibitory factors, 84–86
  overview, 63
Bone mineralization
  calcitriol in, 66–67
  insulin in, 68
Bone morphogenic proteins, in bone cell proliferation, 72
Bone resorption, *see also* Osteoclasts
  calcitonin as inhibitor, 79–80
  calcitriol in, 78–79
  osteoclastic, studies of factors controlling, 77
  parathyroid hormone and, 65–66, 78
  possible mediation by interleukin-6 production, 74
Brain, 14-3-3 proteins
  activation of tryptophan and tyrosine hydroxylases, 153–154
  discovery, 150

C

Calcitonin, 79–80
Calcitriol
  in bone cell proliferation, 66–67
  osteoclast formation stimulation by, 78–79
Calcium, enhanced renal reabsorption
  calcitriol in, 79
  PTH and PTHrp in, 78
cAMP
  alteration of progesterone antagonist pharmacology, 113–114
  in steroidogenesis, 130, 131–132
  transcriptional regulation of steroidogenic enzymes
    cAMP-dependent regulation, 136–142
    cAMP-independent regulation, 136
    developmental–tissue-specific regulation coupling with cAMP-dependent, 142–143
cAMP-dependent protein kinase, in steroidogenesis
  in cAMP-dependent regulation, 130, 132
  phosphorylation of transcription factors by, 139–140
  proof of essential role, 137
  unresolved questions, 144
cAMP response element-binding protein/cAMP response element, steroidogenic enzyme levels and, 136–137
Cardiovascular disease, *see also* Atherosclerosis
  estrogen replacement therapy in risk reduction, 100
β-Carotene
  cigarette smoking and intake of, 54
  UV-induced immunosuppression in elderly and intake of, 48–51
  versus vitamin A as antioxidant, 37
Carotenoids, effects in elderly on immune response, 48–51
Catecholamine biosynthesis, 14-3-3 protein regulation of, 153–154
Cell cycle, 14-3-3 protein regulation of, 156–158
Cells
  oxidation of LDL *in vitro* by, 6–7
  regulation by 14-3-3 proteins, 153–158
  vitamin E deficiency and premature aging, 45
Chaperones, molecular, 14-3-3 proteins as, 167–168
Chimeric PTHrP-negative mice, generation and germline transmission, 180–181
Cholesterol, in steroidogenesis, 131–132, 144
Chondrocytes, in PTHrP-negative mice
  in histological examination of mutant skeleton, 185–187
  possible role of PTHrP in proliferation, 187–190
Cigarette smoking, antioxidants and, 53–55, 56
Colony-stimulating factors, stimulation of osteoclasts by, 81–82

CREB/CRE, see cAMP response element-binding protein/cAMP response element
Cyclic AMP, see cAMP
Cytokines
　enhancing effects of PTH on osteoclasts and bone resorption, 78
　estrogen in regulation, 68, 73
　TGF-β regulation of gene expression, 69
　variation of effects on bone cells and models assessing, 86
Cytotoxicity, oxidized low-density lipoprotein, 4

D

Dehydroascorbic acid
　LDL oxidation modification prevented by, 13
　α-tocopherol oxidation prevented by, 12
Delayed-type hypersensitivity skin test response
　antioxidant supplementation improving, 56
　description, 38–39
　as predictor of morbidity and mortality in elderly, 45
　UV-induced immunosuppression and β-carotene intake, 49–50
　vitamin E supplementation in elderly and enhanced, 52
Deoxyribonucleic acid
　estrogen receptor interaction with antiestrogens and, 117–119
　signal transduction pathway initiation, 106–107
　progesterone receptor interaction with, 112–113
1, 25-Dihydroxyvitamin D_3, see Calcitriol
DTH skin test, see Delayed-type hypersensitivity skin test response

E

Elderly, immune response
　antioxidant effects, 45–46, 46–53
　decline, 44–45
　DTH skin test response, 39, 45
　vaccination responses, 39, 46

Enzymatic activity
　14-3-3 protein, 158
　Pseudomonas aeruginosa toxin, 155–156
Enzymes
　with antioxidant activities, 36
　proteolytic, osteoclastic bone resorption by production of, 77
Estrogen
　in bone cell proliferation, 67–68
　mechanism of action, 101–109
　nonreproductive-related activities, 100–101
Estrogen receptors, see also Ovarian steroid hormone receptors
　antagonism of function, 112–120
　description, 101–103
　initiation of signal transduction pathway, 106–107
　transcriptional regulation by, 103–106
Exocytosis, 14-3-3 proteins in regulation, 154–155
Exoenzyme S, 155–156

F

Fatty acids, see also Polyunsaturated fatty acids
　effects on immune cells, 40
Fatty streaks, in atherosclerosis pathogenesis, 2
Fetal development
　PTHrP-negative mice
　　skeletal abnormalities, 181–184
　　survival until parturition, 181
　　steroid hormones in, 142–143
　　tissue-specific regulation and, 134–135
Fibrinolysis, oxidized LDL as impediment to, 4–5
Fibroblast growth factors, heparin-binding, 72
Foam cells
　conversion of oxidized LDL-derived lipids to, 3
　in pathogenesis of atherosclerosis, 2
　scavenger receptor-independent mechanisms and, 4
14-3-3 proteins
　as molecular bridges, 168
　overview, 149–150

14-3-3 proteins (Continued)
  protein kinases and
    Bcr and Bcr-Abl, 165–166
    protein kinase C, 158–161
    Raf, 161–165
  regulation of cellular processes by
    ADP ribosylation, 155–156
    catecholamine biosynthesis, 153–154
    cell cycle, 156–158
    exocytosis, 154–155
  sequence and structure, 150–153
  unresolved definitive role, 169
Free radicals
  cigarette smoking and, 53, 55
  enhancement of inflammatory response in RA, 43
  immune cell function and
    in aging, 44
    description, 36, 39–42
  LDL and
    ascorbic acid scavenging, 10–12
    transfer by α-tocopherol to, 19–20
  micronutrients interfering with generation, 36–37

G

Gene knockout mice
  consistent female phenotype of SF-1, 143
  PTHrP-negative, method for generating, 179–181; see also PTHrP-negative mice
Genes, see also Transcription; Transcription factors
  expressed in bone development, 64–65
  murine PTHrP
    disruption in embryonic stem cells, 179–180
    perinatal lethality with disruption of, 181
    transgenic technology in expression studies, 177–178
  steroidogenic enzymes
    cAMP reponse sequences for bovine steroid hydroxylase, 138–139
    developmental–tissue-specific regulation, 134–135
Glucocorticoids, in bone cell proliferation, 68
Growth factors
  glucocorticoids and, 68–69
  mediating PTH-stimulated bone formation, 66
  TGF-β regulation of gene expression, 69
  types produced by osteoblasts, 64
Growth hormone, in bone cell proliferation, 68
GTP-binding proteins, 14-3-3 protein and ADP ribosylation of, 155–156

H

Heat shock proteins, see also Chaperones
  ovarian steroid receptors and, 104–106
Heterodimerization, 14-3-3 protein isoforms, 152
Hypercalcemia, induced by PTH and PTHrP, 177

I

ICI 164,384 antiestrogen, 117–119
Immune cells
  β-carotene intake by cigarette smokers and, 54–55
  free radicals and function of, 36, 39–42
  tumor cell lysis
    in elderly, 46, 48
    types capable of, 39
  vitamin E intake by elderly and, 53
Immune response
  assessment of oxidative stress and antioxidant status
    DTH skin text, 38–39
    invasive measures, 38
    noninvasive measures, 37–38, 38
  dependency on free radical and antioxidant status, 55–56
  dysfunction
    in aging, 44–53
    in cigarette smokers, 53–55
    in rheumatoid arthritis, 42–44
  free radicals in depression of, 36
  overview, 37
Insulin, in bone cell proliferation, 68
Insulin-like growth factors I and II, in bone cell proliferation, 70–72
γ-Interferon, as inhibitor of bone resorption, 85

Interleukin-1 effects
 osteoblasts and bone formation, 74
 osteoclasts and bone resorption, 80–81
Interleukin-1 receptor antagonist, 74–75
Interleukin-4
 in bone cell proliferation, 75
 as inhibitor of osteoclast formation and bone resorption, 86
Interleukin-6 effects
 osteoblasts and bone formation, 73–74
 osteoclasts and bone resorption, 83–84
Interleukin-11, in bone cell proliferation, 75

L

LDL, see Low-density lipoprotein
Ligand-binding domains, estrogen and progesterone receptors, 102–103
Lipid hydroperoxides, 13–14
Lipid peroxidation
 lag phase
  extension by α-tocopherol, 16–17, 25
  sequence of antioxidant consumption, 15
 in oxidation of LDL, 5–9
 tocopherol-mediated, 17–20
Low-density lipoprotein
 antioxidant protection by vitamin C, 9–14
 oxidation
  mechanisms, 5–9
  questions for future research, 26
  vitamin C supplementation and, 21–22
  vitamin E and, 14–21, 22–27
 oxidized, 2–5
 in pathogenesis of atherosclerosis, 1–2
Lymphocytes
 carotenoid intake and proliferative response to mitogens, 48
 vitamin E and fish oil intake versus blastogenic response of, 40
 vitamin E supplementation in elderly and enhanced proliferation of, 52
Lymphotoxin, see Tumor necrosis factor-β

M

Macrophages, see also Monocytes
 in lungs of cigarette smokers, 54
 in pathogenesis of atherosclerosis, 2
 prostaglandin secretion in vitamin-E deficient rats, 51–52
 uptake of oxidized LDL, 3
Mammalian cells, 14-3-3 proteins in cell cycle, 157–158
Metal ions
 binding to LDL, ascorbic acid inhibition, 13
 in LDL oxidation, 7–8
Micronutrients, with antioxidant activities, 36
Mitochrondria, in steroidogenesis, 131–132
Molecular bridges, 14-3-3 proteins as, 168
Molecular chaperones, 14-3-3 proteins as, 167–168
Monocytes, see also Macrophages
 in pathogenesis of atherosclerosis, 1–2
 recruitment and retention in atherosclerotic lesions, 3–4

N

Natural killer cells, effects of β-carotene intake
 cell number and function, 50–51
 tumor lysing, 48
Neutrophils
 effects of oxidative burst, 40–41
 in lungs of cigarette smokers, 54

O

Onapristone, 112–113
Osteoblasts
 in bone formation process, 63
 factors involved in proliferation and differentiation
  local, 69–76
  systemic, 65–69
 general characteristics, 64
 systems for in vitro study, 64–65
Osteochondrodysplasia, in PTHrP-negative mice, 181, 184
Osteoclasts
 autocrine–paracrine factors stimulating, 80–84
 general characteristics, 76–77

Osteoclasts (*Continued*)
  hormones affecting function and formation, 78–80
  local inhibitory factors, 84–86
  postmenopausal increase in activity, 100
Osteopenia
  excess glucocorticoids in causing, 68
  insulin deficiency and incidence, 68
Osteoporosis, estrogen deficiency in, 67, 100
Ovarian steroid hormone receptors, *see also* Estrogen receptors; Progesterone receptors
  antagonists, 109–120
  molecular chaperones and, 167
  as transcription factors, 107–109
Ovarian steroid hormones, *see* Estrogen; Progesterone
Oxidants
  free, ascorbic acid scavenging, 10–12
  immune-generated, effects, 40–41
Oxidative stress
  immune response, assessment
    invasive measures, 38
    noninvasive measures, 37–38
  increased, antioxidant status, 55–56
Oxidized low-density lipoprotein, 2–5; *see also* Low-density lipoprotein, oxidation
Oxygen, singlet, micronutrients active against, 36–37

P

$P450_{scc}$ enzyme, in steroidogenesis, 133, 144
Paget's disease
  calcitonin as therapeutic agent in, 79–80
  interleukin-6 and, 83–84
Parathyroid hormone
  osteoblast activity and, 65–66
  osteoclast activity and, 78
Parathyroid hormone–parathyroid hormone-related peptide receptor
  and PTHrP, expression in fetal tibia, 187–190
Parathyroid hormone-related peptide, *see also* PTHrP-negative mice
  in bone cell proliferation, 75–76
  description, 177–179

disorders possibly caused by abnormal levels, 190–191
  and PTH–PTHrP receptor, expression in fetal tibia, 187–190
Peptide hormones
  purpose, 143–144
  in steroidogenesis
    acute action, 130–133, 144
    amplification of regulation of gene transcription, 129–130
    chronic action, 134–143
    future research directions, 144–145
Phenotypic markers, differentiated osteoblasts, 65
Phosphatases, 14-3-3 protein prevention of Raf inactivation by, 162–163
Phospholipase A_2 activity, 14-3-3 protein ζ isoform and, 158
PKA, *see* cAMP-dependent protein kinase
Placenta, versus other producers of steroids, 131
Plaques, atherosclerotic, 2
Platelet-derived growth factor, 72–73
Polyunsaturated fatty acids
  autoxidation, 7
  variations in LDL content, 5
Progesterone
  mechanism of action, 101–109
  nonreproductive-related activities, 100
Progesterone receptors
  antagonism of function, 112–117
  description, 101–103
  initiation of signal transduction pathway, 106
  transcriptional regulation by, 103–106
Prostaglandins
  effects
    osteoblasts and bone formation, 73
    osteoclasts and bone resorption, 79–80
  immunosuppressive, dietary fatty acid intake and synthesis, 40
  secretion levels in vitamin-E deficient rats, 51–52
Protein kinase A, *see* cAMP-dependent protein kinase
Protein kinase C
  binding to 14-3-3 proteins, 152
  phosphorylation of 14-3-3 protein, controversy, 154–155
  regulation by 14-3-3 proteins, 158–161

INDEX

Proteins
  14-3-3 proteins as molecular bridges for, 168
  types produced by osteoblasts, 64
Pseudomonas aeruginosa toxin, 14-3-3 proteins and enzymatic activity, 155–156
PTH, *see* Parathyroid hormone
PTHrP, *see* Parathyroid hormone-related peptide
PTHrP-negative mice
  phenotype
    histological findings in bone tissue, 184–187
    homozygous, perinatal lethality, 181
    PTHrP and PTH–PTHrP receptor expression, 187–190
    skeletal abnormalities, 181–184
  strategy for generating, 179–181
PUFAs, *see* Polyunsaturated fatty acids

R

RA, *see* Rheumatoid arthritis
RACK, *see* Receptors for activated C kinase
Raf protein kinase, 161–165
Receptors
  interaction of hormones with, 99
  ovarian steroid hormone, *see* Estrogen receptors; Ovarian steroid hormone receptors; Progesterone receptors
  scavenger, mediation of macrophage uptake of oxidized LDL, 3
Receptors for activated C kinase, 159
Rheumatoid arthritis
  free radicals and antioxidant intake in, 42–44
  vitamin E and possible pain reduction, 56
RU486 antiprogestin, 112

S

Saka cells, 80
Selenium, antioxidant activity, 36
Serotonin biosynthesis, 14-3-3 protein regulation of, 153–154

SF-1, *see* Steroidogenic factor-1
Signal transduction pathways
  estrogen receptor-mediated, initiation, 106–107
  14-3-3 proteins and, 166–169
  progesterone receptor-mediated, initiation, 106
Singlet oxygen, micronutrients active against, 36–37
Skeletal abnormalities, in PTHrP-negative mice, 181–184
StAR, *see* Steroidogenic acute regulatory protein
Steroid hormones, *see also specific steroid hormones*
  tissues producing, 129, 130–131
Steroid hydroxylases
  bovine, cAMP response sequences, 138–139
  14-3-3 protein interaction with, 153–154
Steroidogenesis, peptide hormones and
  acute action, 130–133
  chronic action, 134–143
Steroidogenic acute regulatory protein, 133, 144
Steroidogenic enzymes
  transcriptional regulation
    cAMP-dependent, 136–142
    cAMP-dependent and developmental and tissue-specific, coupling, 142–143
    cAMP-independent, 136
    developmental and tissue-specific, 134–135
Steroidogenic factor-1
  in cAMP-dependent transcription of bovine steroidogenic enzyme genes, 140, 141
  in developmental expression of steroidogenic enzymes, 134–135, 142–143
Superoxide radicals
  ascorbic acid scavenging, 10–12
  in LDL oxidation, 7–8

T

Testosterone, 142–143
Thrombosis, cause, 4–5

α-Tocopherol, *see also* Vitamin E
  ascorbic acid regeneration, 12
  average LDL particle content, 5, 14
  concentration versus antioxidant protection of LDL, 15–17
  prooxidant versus antioxidant activity, 17–21, 26
  scavenging of lipid peroxyl radicals, 14–15
Transactivation function, ovarian steroid receptors
  description, 108
  estrogen, 103
Transcription
  ovarian steroid receptors and
    description, 107–109
    overview, 101
    regulation by, 103–106
  peptide hormone regulation, *see also* Steroidogenic enzymes, transcriptional regulation
    amplification, 129–130
    pathway from anterior pituitary to all tissues, 130
Transcription factors
  steroid hormone receptors as, 107–109
  in steroidogenesis, phosphorylation by PKA, 139–140
Transforming growth factor-α, 82
Transforming growth factor-β
  as inhibitor of osteoclastic bone resorption, 84–85
  in osteoblast proliferation, 69–70
  production, estrogen stimulation, 67–68
Tryptophan hydroxylase, 153–154
Tumor cells
  lysis
    assessment, 39
    β-carotene intake in elderly and, 48
  TGF-α production, 82
Tumor necrosis factor-α
  in bone cell proliferation, 74
  stimulation of osteoclastic bone resorption, 82–83
Tumor necrosis factor-β, 82–83
Tyrosine hydroxylase, 153–154

U

Ultraviolet light, induced immunosuppression, 48–51

V

Vaccine titers
  immune function indicated by, 39
  vitamin E intake by elderly and, 52
Vasoconstrictors, production, stimulation by oxidized LDL, 4
Vitamin C
  antioxidant protection of LDL by
    in vivo supplementation trials, 21–22
    molecular mechanisms, 10–14
    overview, 9–10, 26
  cigarette smoking and serum levels, 53
  deficiency, immune response in, 41–42
  intake, and immune response in elderly, 53
Vitamin D_3, *see* Calcitriol
Vitamin E, *see also* α-Tocopherol
  cigarette smoking and, 55
  deficiency, premature aging of cells in, 45
  and fish oil intake, lymphocyte blastogenic response, 40
  immune response in elderly and intake, 51–53
  as most critical antioxidant in blood, 36–37
  supplementation
    in atherosclerosis, 22–27
    in rheumatoid arthritis, 44, 56

Y

Yeast, 14-3-3 proteins
  in cell cycle, 156–157
  Raf protein kinase activation, 162

Z

ZK98299, *see* Onapristone

ISBN 0-12-709852-6

90040